The Inner Life of Animals

Love, Grief and Compassion – Surprising
Observations of a Hidden World

PETER WOHLLEBEN

Translated from the German by Jane Billinghurst

THE BODLEY HEAD
LONDON

1 3 5 7 9 10 8 6 4 2

The Bodley Head, an imprint of Vintage Publishing,
20 Vauxhall Bridge Road,
London SW1V 2SA

The Bodley Head is part of the Penguin Random House group of companies
whose addresses can be found at global.penguinrandomhouse.com.

Penguin
Random House
UK

Copyright © Ludwig Verlag, Munich,
part of the Random House GmbH publishing group, 2016

Peter Wohlleben has asserted his right to be identified as the author of this
Work in accordance with the Copyright, Designs and Patents Act 1988

Translation copyright © Jane Billinghurst 2017

Jane Billinghurst has asserted her right to be identified as the translator of this
Work in accordance with the Copyright, Designs and Patents Act 1988

First published in Germany as *Das Seelenleben der Tiere* in 2016

First published in Great Britain by The Bodley Head in 2017

www.penguin.co.uk/vintage

A CIP catalogue record for this book is available from the British Library

ISBN 9781847924551

Typeset in India by Integra Software Services Pvt. Ltd, Pondicherry

Printed and bound by Clays Ltd, St Ives plc

Penguin Random House is committed to a sustainable future for
our business, our readers and our planet. This book is made
from Forest Stewardship Council® certified paper

MIX
Paper from
responsible sources
FSC
www.fsc.org
FSC® C018179

Contents

Introduction   1

Selfless Mother Love   5

Instinct – A Second-Rate Emotion?   13

Loving People   19

Anybody Home?   29

Pig Smarts   39

Gratitude   45

Lies and Deception   51

Stop, Thief!   57

Take Courage!   65

Black and White   71

Cold Hedgehogs, Warm Honey Bees   77

Crowd Intelligence   87

Hidden Agendas   93

Simple Sums   97

Just for Fun   101

Desire   105

Till Death Do Us Part   109

What's in a Name?   113

Grief   121

Shame and Regret   125

Empathy 133

Altruism 139

Upbringing 143

Getting Rid of the Kids 147

Once Wild, Forever Wild 151

Snipe Mess 161

Something Special in the Air 165

Comfort 171

Weathering the Storm 177

Pain 183

Fear 187

High Society 205

Good and Evil 207

Hey, Mr Sandman 215

Animal Oracles 221

Animals Age, Too 229

Alien Worlds 235

Artificial Environments 243

In the Service of Humanity 249

Communication 253

Where Is the Soul? 259

Epilogue 263

Acknowledgements 269

Notes 271

Introduction

ROOSTERS THAT DECEIVE THEIR HENS? Mother deer that grieve? Horses that feel shame? Up until just a few years ago, such ideas would have sounded absurd, mere wishful thinking on the part of animal lovers who wanted to feel closer to their charges. I've been around animals all my life and I was one of those dreamers. Whether it was the chick in my parents' garden that picked me out as its mum, the goat at our forest lodge that brightened our days with her contented bleating or the animals I met on my daily rounds of the woodland that I manage, I often wondered what was going on inside their heads. Is it really true, as scientists have long maintained, that people are the only animals capable of enjoying a full range of emotions? Has Creation really engineered a unique biological path for us? Are we the only ones guaranteed a life of self-awareness and satisfaction?

If that were the case, you wouldn't be reading this book. If human beings were the result of some special biological design, we wouldn't be able to compare ourselves to other animals. It would make no sense to talk about empathy with them, because we would not be able even to begin to imagine how they felt. Luckily, Nature opted for the economy plan. Evolution 'only' modifies and builds on whatever is already available, much like a computer system. And so, just as code from earlier operating systems is integrated into the latest Windows program, the genetic programming of our ancient ancestors still works in us – and in all the other species whose family trees branched off from our lineage in the past few million years. And so, as I see it, there is only one kind of grief, pain or love. It might sound presumptuous to say that a pig feels things just as we do, but there is a vanishingly small chance that an injury hurts a pig less than it hurts us. 'Aha,' the scientists might interject at this point, 'but we have no proof.' That's true, but there never will be any proof. I can't even prove that you feel the same way as I do. No one can look inside another person and prove that, say, the prick of a pin triggers the same sensation in each one of the seven billion people on this planet. But we are able to express our feelings in words, and this ability to share increases the probability that people operate on roughly the same level when it comes to feelings.

So when our dog Maxi polished off a bowl of dumplings in the kitchen and then looked up at us with an innocent expression on her face, she was not behaving like a biological eating machine; she was behaving like the shrewd and endearing little rascal she was. The more often and the more closely I paid attention, the more I noticed our pets and their wild woodland relatives displaying what are supposed to be exclusively human emotions. And I am not alone in this. More and more researchers are realising that humans and many animals have things in common. True love among ravens? No question. Squirrels who know the names of their close relatives? That's been documented for a long time. Wherever you look, animals are out there, loving each other, feeling each other's pain and enjoying each other's company.

Currently there's a great deal of scientific research on the inner lives of animals, although it's usually so narrowly focused and written in such dry, academic language that it hardly makes for gripping reading and, more importantly, rarely leads to a better understanding of the subject. And that's why I would like to act as your interpreter and translate fascinating scientific research into everyday language for you, assemble the individual pieces of the puzzle so that you can see the big picture, and sprinkle in a few observations of my own to bring it all to life. I hope this will help you see the animal world around you, and the species described in this book, not as mindless automatons driven by an inflexible genetic code, but as stalwart souls and lovable rascals. And that is just what they are, as you will discover for yourself when you take a walk in my neighbourhood with my goats, horses and rabbits, or in the parks and woods where you live. Come on. I'll show you what I mean.

Selfless Mother Love

IT WAS A HOT SUMMER DAY at my forester's lodge deep in the woods near Hümmel in the Eifel, a mountain range in Germany. The year was 1996. To cool off, my wife and I had set out a wading pool under a shady tree in the garden. I was sitting in the water with my two children and we were enjoying juicy slices of watermelon when, all of a sudden, I became aware of a movement out of the corner of my eye. A rusty brown something was scampering towards us, freezing for an instant every now and then as it advanced. 'A squirrel!' the children cried in delight. My joy, however, soon turned to deep concern as the squirrel took a few more steps and then keeled over onto its side. It was clearly ill, and after it had taken a few more steps – in our direction! – I noticed a large growth on its neck. It looked like an animal that was suffering from something and might even be highly infectious. Slowly but surely, it was approaching the pool. I was on the point of gathering up the children and beating a hasty retreat when the menacing advance resolved itself into a touching scene. The lump turned out to be a baby squirrel wrapped around its mother's neck like a furry ruff. The baby's stranglehold, along with the shimmering heat, meant that the squirrel mother could only suck in enough air to take a few steps before falling over sideways, exhausted and gasping for breath.

A squirrel mother cares for her children with self-less devotion. When danger threatens, she carries them to safety in the manner I have just described. She can end up totally spent, because she may have as many as six tiny tots to carry, one after the other, each one clasped around her neck. Despite her devotion, the chances of her little ones surviving are low, and about 80 per cent die before they are a year old. Although the rusty rascals can avoid most enemies during the day, death stalks them at night while they are sleeping. When darkness falls, predatory pine martens creep through the branches to interrupt the squirrels' dreams. When the sun shines, the danger comes from agile hawks threading their way through the trees on the lookout for a tasty meal. When a hawk spots a squirrel, a spiral of fear begins. And I mean that literally, for the squirrel tries to escape the hawk by disappearing to the other side of the tree. The hawk banks steeply to follow its prey. In a flash, the squirrel disappears to the other side of the tree again. The hawk follows. Moving at break-neck speed, both animals spiral around the trunk. The nimbler one wins – usually the little mammal.

Winter, however, is more devastating than any predator. To make sure they go into the cold season well prepared, squirrels build dreys. They anchor these spherical nests between branches high up in tree crowns, and fashion two separate exits with their paws so that they can escape any uninvited guests. The nest

is made mostly of small twigs, and the interior is cush-
ioned with soft moss that helps conserve heat and
provides a comfortable place to sleep. Comfortable? Yes,
animals value comfort, too, and squirrels don't like
twigs poking into their backs while they're trying to
sleep any more than we do. A soft moss mattress guar-
antees a restful night.

From my office window, I regularly see squirrels
pulling this soft green material from our lawn and
carrying it high up into the branches. And I see something else, too. As soon as the acorns and beechnuts
tumble to the ground in autumn, squirrels gather these
nutrient-rich packages, carry them a few metres and
bury them. They hide these food caches to ensure they
have food over the winter. Instead of going into true
hibernation, they spend most of their winter days dozing.
In this state of winter lethargy, they use less energy
than usual, but they do not shut down completely like,
say, hedgehogs. Every once in a while, a squirrel wakes
up and gets hungry. Then it slips down the tree and
looks for one of its numerous caches. And it looks, and
looks, and looks. At first it's funny watching the little
animal trying to remember where it has hidden its food.
It burrows a bit here, digs a little there, sitting upright
every once in a while as though taking a break to think.
But that doesn't help. The landscape has changed considerably since the autumn. The trees and shrubs have lost
their leaves, the grass has dried up and, worse, everything might now be covered in cottony-white snow. As
the frantic squirrel continues its search, my heart goes

out to it. Nature is ruthlessly sorting out who will live and who will die. Most of the forgetful squirrels – primarily this year's young – will not live to see spring because they will starve to death. Then I find small clumps of beech trees sprouting in the ancient beech preserves. Baby beeches look like emerald-coloured butterflies fluttering at the ends of slender stalks and they usually grow alone. They gather in clumps only in places where a squirrel has failed to retrieve the nuts it stashed, often because it simply forgot where they were, with the fatal consequences I have just described.

I find the red squirrel to be a prime example of how we sort animals into categories. Their dark button-eyes are adorable, their soft fur is a beautiful reddish colour (there are also some that are brownish-black) and they pose no threat to humans. In spring, young trees sprout from their forgotten food caches, so you could say they help to establish new woodlands. In short, we are kindly disposed towards them. We avoid thinking about their favourite food: baby birds. From my office window at the lodge, I am also privy to their predatory raids. When a squirrel scales a tree in spring, conster-nation reigns in the small colony of fieldfares that raise their young in the old pines along the driveway. The little birds, which are related to thrushes, flutter around the trees, chittering and chattering, trying to drive off the intruder. Squirrels are the birds' deadly enemies, because the little mammals calmly help themselves to one downy chick after another. Even nesting cavities offer the baby birds only limited protection. Armed with long, sharp claws at the end of slender paws, squirrels can fish even supposedly well-protected nestlings out of the tree hollows where they are hiding.

So are squirrels bad or are they good? Neither. A quirk of Nature ensures that they arouse our protective instincts, and so we experience positive emotions when we see them. This has nothing to do with them being good or useful. And on the flip side of the coin, their habit of killing the songbirds we also love doesn't mean they are bad, either. The squirrels are hungry and must feed their young, which depend on nourishing milk from their mother. We would be thrilled if squirrels met their need for protein by gorging themselves on the caterpillars of the cabbage-white butterfly. If they did this, our emotional balance sheet would come out 100 per cent in the squirrels' favour, because these pests are a nuisance in our vegetable gardens. But caterpillars are also young animals, and in this case they grow up to be butterflies. And just because the caterpillars happen to like the plants we have earmarked for our dinner doesn't mean that killing butterfly babies counts as a net benefit for the natural world. The squirrels, meanwhile, are not the slightest bit interested in what we think of them. They are too busy surviving and, while they are at it, making the most of life.

But back to maternal love in these little red scamps. Are they really capable of experiencing this emotion? A love so strong that a squirrel mother places a higher value on the lives of her offspring than she does on her own? Isn't it just a case of a spike in

the hormones coursing through the squirrel's veins
that triggers pre-programmed protective behaviour?
Science has a tendency to reduce biological processes
to involuntary mechanics, and so, before painting such
a dispassionate picture of squirrels and other animals,
let's take a look at maternal love in our own species.
What happens in a human mother's body when she
holds an infant in her arms? Is maternal love innate?
Science would say: yes and no. Maternal love itself is
not innate, but the conditions necessary for developing
this love are.

Shortly before a child is born, the hormone oxytocin
flows through the mother's system, which helps her
develop a strong bond with her child. In addition, large
quantities of endorphins – one of the so-called 'feel-good'
chemicals – are released, which dull pain and reduce
anxiety. This cocktail of hormones remains in the mother's bloodstream after the birth of her child, ensuring
that the baby is welcomed into the world by a mother
who is relaxed and in a positive mood. Nursing stimulates further production of oxytocin, and the mother–
child bond intensifies. The same thing happens in many
animals, including the goats that my family and I keep
at our forest lodge. Goat mothers also produce oxytocin.
A mother goat starts getting acquainted with her kids
when she licks off the mucus that covers her babies after
birth. The clean-up process intensifies their bond, and
as the mother goat bleats softly to her children, her
offspring reply in thin, reedy voices and the vocal signa-
tures are imprinted in both mother and kids.

Things do not go well if something goes awry at clean-up time. When a mother goat in our small herd is ready to give birth, we put her in a pen of her own so that she can deliver her kids in peace. There is a small gap under the door of the stall, and once during a birth a particularly small kid slipped out under it. By the time we noticed the mishap, precious time had passed and the mucus covering the kid had already dried. The result? Despite our best efforts, the mother goat refused to accept her baby. The time to trigger mother love had passed.

Something similar can happen with people. If a mother in hospital is separated from her newborn baby for an extended period of time, the maternal bond becomes more difficult to establish. The situation is not as dramatic as it is with goats, because humans are not totally dependent on hormones and can learn how to love. If we were like goats, adoptions would never work out, because adoptive mothers often meet their children years after their birth. Adoption, therefore, is the best opportunity we have for investigating whether maternal love is more than just an instinctive reflex and something that can be learned. But before we tackle this question, I would like to shine some light on instincts and how they work.

Instinct – A Second-Rate Emotion?

I OFTEN HEAR that there's no point comparing animal emotions to human emotions, because animals act and feel instinctively, whereas humans act consciously. Before we turn to the question of whether instinctive behaviour is second-rate, let's take a closer look at instincts. Science uses the term 'instinctive behaviour' to describe actions that are carried out unconsciously, without being subjected to any thought processes. These actions can be genetically hard-wired or they can be learned. What is common to all of them is that they happen very quickly because they bypass cognitive processes in the brain. Often these actions are the result of hormones released at certain times (in moments of anger, for example), which then trigger physical responses. So are animals nothing more than biological automatons on autopilot?

Before rushing to judgement, let's consider our own species. We are not free of instinctive behaviour ourselves. Quite the opposite, in fact. Think about a hot element on an electric stove. If you were to absent-mindedly put your hand on one, you'd take it away again in a flash. There's no preceding conscious reflection, no internal monologue along the lines of 'That's strange. It smells like someone's barbecuing something and my hand suddenly really hurts. I'd better remove it.' You just react automatically without making a conscious decision to remove your hand. So people behave instinctively, too. The question is simply the extent to which instincts determine what we do every day.

To shed some light on the matter, let's turn to recent studies of the brain. The Max Planck Institute in Leipzig published the results of an astonishing study carried out in 2008. With the help of magnetic resonance imaging (MRI), which translates brain activity into digital images, test subjects were observed making decisions: whether to push the computer button with their right hand or with their left. The activity in their brains clearly showed what their choices were going to be, up to seven seconds before the test subjects themselves were aware of them. This means that the behaviour had already been initiated while the volunteers were still considering what to do. And so it follows that it was the unconscious part of the brain that triggered the action. It seems that what the conscious part of the brain did was to come up with an explanation for the action a few seconds later.[1]

Research into these kinds of processes is still very new, and so it's impossible to say what percentage and what kinds of decisions work this way, or whether we're capable of rejecting processes set in motion unconsciously. But still, it's amazing to think that so-called free will is often playing catch-up. All the conscious part of the brain is doing in this case is coming up with a face-saving explanation for our fragile ego, which, thanks to this reassurance, feels it's completely in control at all times. In many cases, however, the other side – our unconscious – is in charge of operations.

In the end, it doesn't really matter how much our intellect is consciously in control. Despite the fact that a surprising number of our reactions are probably instinctive, our experiences of fear and grief, joy and happiness are not at all diminished by being triggered instinctively, instead of being actively instigated. Their origin doesn't reduce their intensity in any way. The point is that emotions are the language of the unconscious and, in day-to-day life, they prevent us from sinking beneath an overwhelming flood of information. The pain in your hand when you put it on a hot element allows you to react immediately. Feeling happy reinforces positive behaviours. Fear saves you from embarking on a course of action that could be dangerous. Only the relatively few problems that actually can, and should, be solved by thinking them through make it to the conscious level of our brain, where they can be analysed at leisure.

Basically then, emotions are linked to the unconscious part of the brain, not the conscious part. If animals lacked consciousness, all that would mean is that they would be unable to have thoughts. But every species of animal experiences unconscious brain activity, and because this activity directs how the animal interacts with the world, every animal necessarily has emotions. Therefore instinctive maternal love cannot be second-rate, because no other kind of maternal love exists. The only difference between animals and people is that we can consciously activate maternal love (and other emotions); for example, in the case of adoption, where there can be no question of an instinctive bond created between mother and child at birth because first contact often happens much later on. Yet here, too, instinctive maternal love develops over time and the accompanying hormone cocktail flows through the mother's bloodstream.

Aha! Have we finally successfully isolated a human emotional domain that animals cannot enter? Let's take another look at our red squirrel. Canadian researchers have been watching its relatives in the Yukon for more than twenty years. About 7,000 animals took part in the study and, although red squirrels are solitary animals, five adoptions were observed. Admittedly, each case involved squirrel babies of a close family member being raised by another female. Only nieces, nephews or grandchildren were adopted, which shows that squirrel altruism has its limits. From a purely evolutionary standpoint, there are advantages to this arrangement, because it means very closely related genetic material is preserved and handed down, although it has to be said that five cases in twenty years is not exactly overwhelming proof of an adoption-friendly attitude in squirrels.[2] So let's take a look at some other species.

What about dogs? In 2012 a French bulldog called Baby hit the headlines. Baby lived in an animal sanctuary in Brandenburg, Germany. One day, six baby wild boar were brought in. The sow had probably been shot by hunters, and the tiny striped piglets wouldn't have stood a chance on their own. At the sanctuary the animals

got full-fat milk – and full-on love. The milk came from
the carers' bottles, but the love and warmth came from
Baby. The bulldog adopted the whole crew right away
and allowed the piglets to sleep snuggled up to her. She
also kept a watchful eye on the little tykes during the
day.[3] But could that be called a true adoption? After all,
Baby didn't nurse the piglets. But nursing is not a necessary component of human adoptions, either, and yet
there are reports of dogs who even did that. A Cuban
dog, Yeti, had just given birth to a litter of puppies,
which meant she had a lot of milk. When a few pigs on
the farm also had babies, Yeti lost no time adopting
fourteen piglets, even though their own mothers were
still around. The little piglets followed their new mum
around the farmyard and, of most importance here, Yeti
nursed them.[4] Was that an example of conscious adoption? Or did Yeti just have maternal instincts to spare?
We could ask these same questions of human adoptions,
where people with strong desires look for and find an
outlet for them. You could even liken the keeping of
dogs and other pets to interspecies adoption; after all,
some four-legged friends are accepted into human
society almost as though they are members of the family.

    There are other cases, however, where super-
abundant hormones or surplus milk can be ruled out as
the driving forces behind adoption. The crow called
Moses is a touching example. When birds lose their
brood, Nature gives them another opportunity to work
off their pent-up impulses. They can simply start anew
and lay another clutch of eggs. There's no way a single

bird like Moses can exercise its maternal instincts, yet Moses attempted to do just this. The target of Moses' attention was a potential enemy – a house-cat, albeit an extremely small and relatively helpless one, because the kitten had obviously lost its mother and had not had anything to eat in a long time. The little stray popped up in Ann and Wally Collito's garden. The couple lived in a cottage in North Attleborough, Massachusetts, and they watched in amazement at what happened next. The crow attached itself to the little orphan and was clearly looking after it, feeding it with earthworms and beetles. Of course the Collitos didn't just stand by and watch; they fed the kitten, as well. The friendship between the crow and the cat continued after the cat grew up, and it lasted until Moses disappeared five years later.[5]

But let's get back to instincts. In my opinion, it makes no difference whether a mother's love is triggered by unconscious commands or comes after conscious deliberation. At the end of the day, it feels just the same. What is clear is that people are capable of both, although instinctive love triggered by hormones is more common. Even if animals are not capable of consciously developing maternal feelings – and the adoption of animals across species barriers should make us rethink that one – instinctive maternal love remains, and it is just as moving and just as compelling. The squirrel that crossed our lawn in a haze of heat with a baby wrapped around her neck was motivated by deep devotion. And, when I think back on that day, knowing that makes the experience all the more beautiful.

Loving People

CAN ANIMALS REALLY LOVE US? We've already seen, in the case of squirrels, just how difficult it can be to verify this feeling between animals of the same species. But to now add love across the species divide – and all the way to us humans? You've got to wonder whether this is simply wishful thinking to make it easier for us to justify imprisoning our pets. First, let's take another look at the mother–child bond, because this particularly strong kind of love is something we can actually trigger in animals, as I experienced when I was a boy.

Even back then, my interests revolved around nature and the environment, and I spent every spare moment outside in the woods or at lakes in abandoned quarry pits along the Rhine. I imitated the calls of frogs to get them to respond, kept a few spiders in glass jars so that I could observe them, and raised mealworms in flour to watch them turn into black beetles. In the evening I curled up with books about behavioural biology (don't worry, books by adventure writers such as Karl May and Jack London also had their place by my bed). In one of them, I read that you can get chicks to imprint on people. All you had to do was incubate an egg and talk to it just before it hatched, so that the tiny creature inside became imprinted on a human instead of a hen. Apparently, the relationship lasted a lifetime. How exciting!

At the time, my father kept a few hens and a rooster in the garden, so I had access to fertilised eggs. I didn't have an incubator, so an old electric blanket had to do. There was one other problem. Chicken eggs need to be kept at 38 degrees Celsius and turned often every day, so that they can cool down a bit. Armed with a scarf and a thermometer, I had to painstakingly simulate behaviour that comes naturally to a hen. For twenty-one days I measured the egg's temperature, draped varying layers of scarf over it and carefully turned it. A few days before the estimated date of hatching, I began my monologues. And then it happened. Punctually on the twenty-first day a small packet of fluff pecked its way to freedom. I immediately christened it Robin Hood.

The chick was incredibly adorable. Its yellow feathers were sprinkled with tiny black spots and its black button-eyes gazed straight at me. It never wanted to leave my side, and every time it lost sight of me, it began to cheep frantically. It didn't matter if I was on the toilet, in front of the television or in bed, Robin was always there. The only time I left the little chick alone was when I went to school. Then I took my leave with a heavy heart, and I was greeted effusively when I returned. But this intimate bond began to stress me out. My brother took pity on me and cared for the chick part-time so that I could do something without Robin every once in a while. Eventually, however, it became too much for him as well. By now, Robin had developed

into a young hen, and we gave it to a retired English
teacher who was very fond of animals. Man and hen
became fast friends, and for a long time you could see
the two of them taking walks together in the neighbouring village: the teacher on foot, with Robin riding
on his shoulder.

I think it's safe to say that Robin established a
genuine relationship with its human carers, and many
people can share similar stories about being a substitute
parent for a young animal. The bottle-fed kids my wife
hand-raises, for example, remain extremely attached to
her for life. Here and in other cases, human carers play
the role of adoptive mothers, and the stories are always
heart-warming. However, these relationships are not
voluntary – at least not as far as the animals are
concerned, even if they have to thank their carers for
their survival. It would be more meaningful if an animal
were to come and stay with us of its own accord. But
has this ever happened?

To find out, we must leave the warm embrace of
maternal love and cast a wider net. What we're looking
for is a scenario where an animal can grow up and decide
for itself whether it will stay or leave. There's a good
reason most dogs and cats come to us as babies, because
that removes the element of choice for the little scamps.
And that's absolutely a good thing. After a few days of
getting used to their new circumstances – and possibly
after a twinge of anxiety at being separated from their
mother – young animals just a few weeks old quickly
get attached to their carers and, exactly like my wife's

bottle-fed kids, they remain particularly close to them for as long as they live. Everyone feels good, but there's still that nagging question: are there any adult animals that enter into relationships with people of their own free will?

For house-pets, the answer is a resounding yes. There are countless examples of stray cats and dogs that practically force themselves on caring humans. But in answering this question, I'd prefer to explore the world of wild animals, because wild animals have not had tameness bred into them and are therefore not predisposed to seeking a connection with people. And I'd like to exclude one more scenario: using food to tame animals. When wild animals are offered food, the only thing they want to do is eat and therefore they tolerate, and to a certain extent get used to, our presence. Our former neighbours found out what a nuisance this can be when they started feeding a squirrel. For weeks they had been tempting the little rascal with nuts, and it had practically become a member of the family. But if the human food dispenser wasn't there in a timely manner every day, the squirrel would start scratching impatiently at the window. It demolished the frame in just a few weeks; and squirrel claws are razor-sharp.

Most friendships between wild animals and people are to be found in the sea, with dolphins. Fungie, who lives in Dingle Bay in Ireland, is a particular star. He pops up often, accompanies tour boats and shows off for visitors. He's become a real tourist magnet and features in official travel brochures. People who feel moved to do so can safely get into the water with him. The sizeable dolphin swims alongside them, and they experience a special kind of joy in his presence. His tameness doesn't depend on food, which he refuses to accept. Fungie has been around for more than thirty years now, and it's difficult to imagine life in Dingle without him. Most people find his story delightful – but not everyone. A reporter for the German newspaper *Die Welt* interviewed scientists and asked whether the dolphin might not simply be deranged. Perhaps, the reporter asked, the solitary animal hangs out with people only because he's shunned by others of his kind?[6]

Apart from the fact that people often form friendships with animals for similar reasons – for example, because they are lonely after the loss of a partner – I would like to investigate the question further with land-based animals closer to home. And that's not easy, because a common characteristic of wild animals is that they are exactly that – wild – and therefore they normally don't seek contact with people. Moreover, people have hunted them for tens of thousands of years, so they have evolved to be wary of humans; those that don't escape in time are in danger of losing their lives. And that is still the case for many animals, as you can see just by running your eye down the list of animals that it's still legal to hunt. Whether they are large game such as deer or wild boar, or smaller four-footed targets such as foxes or hares, or even birds, from raptors to geese and ducks or snipe, every year thousands upon thousands meet their end in a hail of bullets. Thus a certain mistrust on the part of anything on two legs is completely understandable. And that is why we are so moved when such a creature overcomes its natural wariness and seeks contact with us.

What might motivate a wild animal to do that? Let's dismiss attracting them with food, because then we don't know whether it's just a case of hunger overriding fear. There is another driving force, however – one that is important for people as well – and that is curiosity. My wife, Miriam, and I had the good fortune to encounter at least one curious species: reindeer in Lapland. Okay, the reindeer are not completely wild, because the indigenous people, the Sami, own the animals and herd them with helicopters and all-terrain vehicles when they want to sort them for butchering or branding. Yet despite this, the reindeer have retained their wild character and are usually very wary around people.

Miriam and I were camping in the mountains in Sarek National Park, and because I am an early riser by nature, I was the first to creep out of my sleeping bag in the morning. I had been gazing for a while at the breathtaking sight of Nature untouched by human hands when I suddenly became aware of movement close by. A reindeer! Just the one? No, there were more coming down the slope, and I woke Miriam so that she could watch the animals as well. As we ate breakfast, more and more reindeer gathered round, until we were surrounded by the whole herd – about 300 animals. The reindeer spent all day around our tent, and one young calf even dared to get within a few metres, so that it could lie down by the tent for a midday nap. We felt we were in paradise.

When a small group of hikers walked by, we realised how wary of people these animals really were. As soon as the hikers appeared, the herd retreated, only to return a while later to the area around our tent. It was clear that some of them were very interested in us. Eyes wide open and nostrils flared, they tried to figure us out. For us, it was the most amazing experience of the whole trip. We had no idea why the reindeer were so trusting around us. Perhaps our body language was calmer than usual for humans, because of our daily interactions with animals, and that made us seem less threatening.

Anyone can have similar interactions in places where animals are not hunted. In national parks in Africa, for instance, or on the Galapagos Islands, or out on the tundra in the far north – places where species have not yet had bad experiences with people – animals allow visitors to get very close to them. And every once in a while there are some individuals that are curious enough to want to check out the unusual guests wandering about in their territory. These are the encounters that make people particularly happy, because here both parties meet completely voluntarily.

It is difficult to prove that an animal truly loves a person of its own free will. Even my little chick, Robin Hood, had no real choice but to develop feelings for me. How about looking at the other side? Every owner of a pet, be it a cat or a dog or some other animal, knows that people are capable of loving animals. But what about the quality of this love? Some might argue that people simply project their emotions onto animals and see them reflected back. Their pets are substitutes for children they wished they'd had, partners they've lost or friends who keep their distance. The subject is a minefield that I'd much rather avoid; however, as we're talking about animal emotions, we should ask how our sentimental attachments affect our four-legged friends.

First off, they literally deform animals. In most places in the world it's been a long time since cats and dogs were bred to be highly skilled helpers in hunting hares, deer or mice. Instead, we've been breeding them to satisfy, in both character and appearance, our desire to have something to cuddle and hug. French bulldogs are a good example. I used to think they were ugly, and that their squashed, wrinkled snouts put them at a disadvantage, because their snub noses interfere with their breathing so much that they snore. But then I got to know Crusty, a blue-grey male that we looked after every once in a while. Crusty won me over right away and, from that moment on, I no longer cared how he had been bred – he was just so adorable. Whereas other dogs have had enough after five minutes of being stroked, Crusty enjoyed this treatment for hours. If you stopped, he would nudge your hand beseechingly and look at you with his big puppy-dog eyes. His favourite activity was snoring contentedly while sleeping on his owner's stomach.

Can breeding like that be a bad thing for the dog? There's no question that French bulldogs have been

bred to be lapdogs – living cuddly toys, so to speak. I
don't want to judge the legitimacy of this. The more
important question I'd like to ask is: what is this like
for the dog? If a heightened need to be stroked has been
bred into it, and if its appearance causes everyone (and
I mean everyone!) to want to satisfy this need immediately, does the dog have a problem? It obviously feels
just fine, and both it and the people it meets get what
they wish for. It's just that what led to this need to be
stroked – genetic manipulation through selective
breeding – has a tiny trace of the unnatural about it.
This is very different from cases where owners ignore
their animals' needs, whether natural or caused by
breeding, and when self-interested love blinds them so
much that they end up treating their pets like people
dressed in dog costumes. In such cases, overfeeding,
insufficient exercise and lack of exposure to the delights
of the outdoors (such as walks in the snow) lead to
severe health problems that torture the pampered
animals to death.

Anybody Home?

BEFORE WE DELVE DEEPER into the emotional and inner life of animals, we should ask once again whether the idea is just too far-fetched. After all, we need to have certain brain structures in order to process the emotions we experience, at least according to current thinking in science. The answer is pretty clear: in people, it is the limbic system that allows us to experience the full range of joy, grief, fear or desire; and together with other areas of the brain, it facilitates the appropriate physical reactions.[7] These brain structures are very old in evolutionary terms and so we share them with many mammals: goats, dogs, horses, cows, pigs — the list goes on and on. According to recent research, not only mammals, but also birds and even fish, which biologists rank far lower on the evolutionary scale, belong on this list.

In the case of aquatic animals, it was research into pain that led to the topic of emotions. The starting point was fishing and whether fish can feel the injuries caused by hooks. What might appear self-evident to you was thought for a long time to be unlikely. When you see photographs of fishing trawlers pulling aboard nets filled with living, slowly suffocating ocean inhabitants, when you see a trout thrashing around at the end of an angler's bent rod, you have to ask yourself how society tolerates such behaviour, in light of today's discussions about animal welfare. It's probably not a case of intentional ill will, but rather acceptance of the mostly unproven assumption that fish are witless creatures that swim around in rivers and oceans, not feeling anything at all.

Victoria Braithwaite, a professor at Penn State who earned her doctorate from the University of Oxford, discovered something quite different. Years ago she identified more than twenty pain receptors around the mouths of fish, right where anglers' hooks usually catch hold.[8] Ouch! But all that proves is that a dull feeling of pain is within the realm of possibility. And so Braithwaite poked needles into the areas she had identified, which caused a reaction in the hindbrain of the fish, which is exactly where pain stimuli are processed in people. That should be enough to prove that fishing injuries cause fish to suffer.

But what about emotions? Let's consider fear. In humans, fear arises in the almond-shaped amygdala, located in the brain's temporal lobes. For a long time that was not proven, even though it was suspected. It wasn't until January 2011 that scientists from the University of Iowa published a paper about a woman identified as SM, who was afraid of spiders and snakes – until the cells in her amygdala died after a rare illness. That was, of course, tragic for SM; however, it offered researchers a unique opportunity to investigate what happens when this organ is lost. They took SM to a pet shop and confronted her with the animals she feared.

The woman could now touch the animals, something she could never have done before, and reported that she simply felt curious about them and no longer felt in the least bit frightened.[9] And so the seat of fear can now be precisely located in people. But what about fish?

Manuel Portavella García and his team at the University of Seville have indeed found comparable structures in the outer areas of the fish brain, a place where no one had looked before. (In people, the fear centre lies deep inside the lower part of the brain.) First, the researchers trained goldfish to swim quickly away from one corner of their tank as soon as a green light came on. If they failed to do this, they got an electric shock. Then the researchers disabled a part of the fish brain known as the telencephalon. It corresponds with our fear centre, and switching it off had the same result as in people: from then on, the goldfish were no longer afraid of the green light and they ignored it. From this, the researchers concluded that fish and terrestrial vertebrates have inherited the same brain structures from common ancestors that lived more than 400 million years ago.[10]

It follows that all vertebrates have had the hardware for emotions for a very long time, but does that mean that animals feel things the same way we do? Much points in that direction. Scientists have even found oxytocin in fish – the hormone that not only brings joy to mothers, but also strengthens the love between partners. Joy and love in fish? We're not going to be able to prove that, at least not in the foreseeable future, but if there's any doubt, why don't we adapt the principle of 'innocent until proven guilty'? Scientists have spoken out against emotions in animals for so long that their view is mostly accepted, but wouldn't it be better to give animals the benefit of the doubt, to be sure that they are not suffering unnecessarily?

In the preceding chapters I purposely described emotions as we experience them. This is the only way we might be able to understand what's going on inside animals' heads. But even if the structures in their brains differ from ours and these differences mean that they probably experience things differently, that certainly doesn't mean that emotions in animals are inherently impossible. It simply means that it is more difficult for us to imagine what their emotions might feel like. Take the fruit fly, for example, whose central nervous system is made up of 250,000 cells, making it 1/400,000th the size of ours. Can such minute creatures, with such a limited capacity up top, really feel anything? Can they even be said to possess consciousness – this being, of course, the pinnacle of achievement? Unfortunately science is not yet advanced enough to be able to answer this question, partly because the concept of consciousness cannot be clearly defined.

The closest we can get to a definition is that consciousness involves thinking and reflecting on things that we have experienced or read about. Right now, you're thinking about what you're reading, and so you possess consciousness. And at a very basic level, the conditions necessary for consciousness have been discovered in the tiny brains of fruit flies. The flies are constantly barraged by stimuli from the external world, just as we are. The smell of roses, car exhausts, sunlight, a breath of air – all are registered by a variety of unconnected nerve cells. So how does a fly filter out from this flood of sensations what is most important, so that it can stay out of danger and not miss out on a particularly tasty morsel? Its brain processes the information and ensures that different areas coordinate their activities, strengthening certain stimuli. And so what is of interest stands out from the general noise of thousands of other impressions. The flies, therefore, can focus their attention on specific things – just as we can.

Fruit-fly eyes are made up of about 600 individual facets. Because these tiny insects dart around so quickly, their eyes are bombarded by a huge number of images every second. This seems like an impossibly large amount of data to process, but the flies must do this if they are to survive. Anything that moves could belong to a voracious predator. Therefore the fruit-fly brain leaves all static images blurry and focuses exclusively on moving objects. You could say that the tiddlers are stripping things down to the bare essentials, an ability that you surely would not have expected these little flies to have. By the way, we do something similar. Our brains don't allow all the images we see to make it through to our consciousness. They only let the important ones through. Does that mean flies have consciousness? Researchers won't go that far; however, it is clear that flies can at least actively focus their attention on what matters most to them.[11]

Let's return to variations in brain structure between species. The basic organ is certainly present even in lower vertebrates, but for the quality of feelings we experience, more is needed. You read over and over again that intense emotions of which the subject is consciously aware are only possible with central nervous systems like ours. The stress lies on the words 'consciously aware'. The grooves and ridges in our brain occur in its outermost layer, the neocortex, which is the most recent part of our brain to evolve. This is the seat of self-awareness and consciousness, the place where thinking happens. And the human brain has more of these cells than any other species. The crowning achievement of Creation sits there right under our skull. It follows that all other creatures must be less aware of the constellation of emotions that we experience and cannot be as intelligent as we are, right? Consider comments made by Germany's first professor of fisheries and fishing, Robert Arlinghaus, co-author of a study on pain in fish for the German government. In an interview with the German magazine *Spiegel Online*, he stressed that fish cannot experience pain as we do, from the injuries they suffer when they are caught, because they do not have a neocortex and therefore they can have no conscious awareness of pain.[12] Apart from the fact that other scientists do not agree with him (see below), this sounds to me more like a rationalisation of his hobby than a reasoned, objective scientific opinion.

Gourmets advance a similar argument every year at Christmas when it's customary to load the table with tasty crustaceans, and *Der Spiegel* (the print sister-magazine to *Spiegel Online*) has reported on this as well.[13] The poster-child for the spectrum of shellfish is the lobster, which is served up on a platter as a status symbol after being boiled bright red. Boiled alive, that is. Whereas vertebrates are killed before they are cooked, it's perfectly acceptable to throw crustaceans into a bubbling pot with all their senses intact. It can take minutes until the heat makes its way completely inside the animal, destroying its sensitive nerve endings. Pain? How can that be? Crustaceans don't have a spinal cord, and that means they are incapable of feeling pain. Or at least that is what people say. Crustaceans' nervous system is configured differently from ours, and it's even more difficult to prove pain in crustaceans than it is in species that have an internal skeleton, like we do. Scientists arguing on behalf of the food industry insist that the animals' reactions are nothing more than reflexes.

Professor Robert Elwood at Queen's University Belfast disagrees. 'Denying that crabs feel pain because they don't have the same biology is like denying they can see because they don't have a visual cortex [the part of the brain responsible for sight in people].'[14] Apart from that, pain is a component in reflex actions, as you can easily test for yourself if there's an electric fence nearby. If you put your hand on it and get an electric shock, you have no choice but to pull back right away, whether you want to or not. It's pure reflex on your part, something you do without a moment's thought, but that doesn't make the electric shock any less painful.

Is there really only one way — the human way — to experience feelings intensely and perhaps consciously? Evolution is not the single-track process we sometimes think — or maybe hope — it is. Birds, some of which possess diminutive brains, are a prime example that there is more than one route to intelligence. Since the age of their ancestors, the dinosaurs, birds' development has followed a different path from ours. Even without a neocortex, they can perform mental feats of the highest order. In birds, a region called the dorsal ventricular ridge oversees similar tasks and functions as our cerebral cortex does. In contrast to the human neocortex, which is built up of layers, the equivalent area in the bird brain is made up of small clumps, a fact that fed long-standing doubt that it could perform a similar function.[15] Today we know that ravens and other species that live in social groups can match, and in some cases even exceed, the mental prowess of primates. This is further proof of science's practice of arguing too cautiously when in doubt about feelings in animals, denying them many mental capacities until there is positive proof that they possess them. Instead, couldn't we simply (and just as accurately) say: we don't know?

Before I end this chapter, I would like to introduce you to one more creature in our woods, an organism that is mindless in the truest sense of the word.

Sometimes you can find it on rotting wood, where it forms a small, bumpy yellow mat. It's a fungus. Hold on a moment. Isn't this a book about animals? Well, in the case of this fungus, science is not exactly sure which category it belongs to. It's difficult enough with normal fungi, which form a third kingdom of living things in between animals and plants, because they cannot be clearly assigned to either category. Like animals, fungi subsist on organic substances from other living beings. In addition, their cell walls are made of chitin, like the exoskeletons of insects. And the slime mould that creates that yellow mat on dead wood can even move. At night, like gelatinous jellyfish, these organisms are capable of slithering out of the glass lab containers where they are temporarily confined. Today, science is moving them out of the realm of fungi and edging them a step closer to animals. Welcome to this book!

Researchers find some kinds of slime moulds so interesting that they regularly observe them in the laboratory. *Physarum polycephalum*, to give its somewhat awkward Latin name, is just such a customer, and it loves rolled oats. Basically, the creature is one giant cell with countless nuclei. Researchers now place these slimy unicellular organisms in a maze with two exits and put food at one of the exits as a reward. The slime mould spreads out into the maze and, after 100 hours or more, finds the exit with the oats. Not bad, really. To do this, it clearly uses its own slime trail to recognise where it has already been. It then avoids those areas because they have not led to success. In Nature, such behaviour is of practical benefit, because the creature knows where it has already been in its search for food and, therefore, the places where there isn't any food left. It's quite a feat to be able to solve a maze when you don't have a brain, and researchers credit these moving mat-like creatures with having some kind of spatial memory.[16] Japanese researchers topped it all off by using a slime mould to reproduce a map of the most important transportation routes in Tokyo. To do this, they set a slime mould down on a damp surface at a point that represented the centre of the city. Piles of food marked the principal neighbourhoods as attractive places to visit. The slime mould set off and when it had connected the neighbourhoods using the optimal, shortest route, there was a big surprise: the image pretty much corresponded to the suburban train system in the metropolis.[17]

I particularly like the slime-mould example, because it shows how little it takes to overturn our preconceptions about primitive Nature and stupid, emotionless animals. These alien creatures lack any of the basics laid out in the preceding chapters and yet, if organisms with only a single cell have spatial memory and can perform complex tasks, how many undreamed-of skills and emotions might there be in animals with as many as 250,000 brain cells, like the fruit flies I've just introduced you to? Given how much more like us birds and mammals are in the physical structure of their bodies and brains, it would hardly come as a surprise if we were to discover that they were as sensitive to the world as we are.

Pig Smarts

DOMESTIC PIGS ARE DESCENDED from wild boar, which were prized by our ancestors as a source of meat. About 10,000 years ago wild boar were tamed to ensure the delicious animals were available at short notice without us having to go out on dangerous hunts to get them, and they were then bred to better satisfy our requirements. Despite this interference, modern domestic pigs have retained wild boar's behavioural repertoire and, above all, their intelligence.

First, let's look at how wild boar behave. (Feral hogs or swine in the US are descended from pigs that escaped domesticity and share behaviours with European wild boar.)[18] For example, wild boar know exactly which other boar they are related to, even if the connection is a distant one. Researchers from Dresden University of Technology determined this indirectly when they were investigating the home ranges of family groups (known as sounders). As part of this research, 152 wild boar were caught in traps or stunned with tranquilliser guns, fitted with transmitters and then set free again so that the researchers could see where these nocturnal roamers hung out. They discovered that there is normally not much overlap between neighbouring sounder territories and, on average, territories range from 4 to 5 square kilometres in size, which is much smaller than previously thought.[19]

Wild boar rub against trees to mark their territorial boundaries. After wallowing in mud, they cover these 'rubbing trees' with their scent. Scent markings, however, are not permanent, which means that boundaries between territories remain somewhat fluid, and so it's little wonder that every once in a while boar intrude where they don't belong. As meeting up with strangers usually leads to altercations that even boar prefer to avoid, border violations by unrelated sounders are fairly rare, but if the home ranges of two related groups are side by side, their territories may overlap by as much as 50 per cent. Clearly, wild boar are more kindly disposed towards family members than they are towards strangers and, most importantly, they can obviously tell the difference.

Family dispersal starts when the previous year's piglets, the yearlings, are driven off as the birth of the next litter approaches. The sow has no extra time to look after older piglets, which are pretty independent by then. Wild boar are highly social and love to engage in mutual grooming or simply to lie snuggled closely together, and the siblings join up to form yearling sounders so that they can continue living in a group. If the yearling sounders run into their extended families with their new piglets later in the year, the meeting will be a friendly one. Everyone knows everyone else and they all still get along well.

Thinking about our domestic animals, I've often wondered whether our goats and rabbits are capable of picking their grown-up children out of the group and recognising them as relations. After observing them for a long time, I believe I can now answer this question with a resounding yes. With one proviso: that the animals are not separated from one another. If they are kept in separate enclosures for more than a few days, they end up treating each other as strangers. Perhaps their long-term memory is not configured for storing information about family relationships. It is clearly different for wild boar, and therefore also probably for domestic pigs, because they have long memories about who belongs to whom. This is of little use to domestic pigs, of course, because unfortunately for them, they are separated from their parents and raised in groups of other pigs their own age, and as a rule, they don't make it past their first year.

These days, most people are aware that pigs are extremely clean animals. They prefer to use some kind of a toilet – a designated place where they do their business – be it big or small. This toilet is never in their sleeping hollow. After all, who would want to sleep in a stinky bed? This goes for both wild and domestic pigs. When you see photographs of tiny stalls in factory-farm barns (allowing 1 square metre per animal) and pigs covered in muck, you can imagine how uncomfortable the animals must be. In the wild, boar adapt their sleeping quarters according to the weather and time of year. Because they select their resting places with care, they prefer to use the same spot all the time; however, when storms blow and rains soak their slumbers, they will move to wooded areas where they can sleep protected from the wind and stay relatively dry. In summer, the bare earth is sufficient as a place to lay their head, because at that time of year wild boar are usually too warm anyway. In winter, however, they plan their nightly repose especially carefully. A snug little nest under a thick windbreak of blackberry bushes, with only two or three tunnel-like entrances, is ideal. They bring in dried grass and leaves, moss and other soft materials, which they pile up carefully to make a cosy bed.

Did I say 'nightly repose'? Although they would probably love to sleep as we do when we lie dreaming in our beds, these smart animals have adjusted their circadian rhythms. Every year in Germany, hunters shoot as many as 650,000 wild boar,[20] and to do that they need daylight. In order to avoid their pursuers, the boar go about their business under the cover of darkness. Normally that would be protection enough, because in Germany it is illegal to shoot animals after dark; however, an exception has been made for wild boar to try to control their burgeoning populations. Because night-vision devices are still prohibited, hunters have to wait for a full moon and clear skies so that they can see more than just vague shadows in woodland clearings. They attract the wild boar with small amounts of feed corn, which the animals are particularly partial to. The goal: to dispatch the boar with a deadly shot while they are feeding. But it's not that simple to outwit these canny animals, which simply put off feeding until the wee hours of the morning. But the hunting industry has a solution at the ready for this, as well: game clocks that stop when they are disturbed. When hunters put these clocks in among the corn, they show the time when the wild boar come to feed. Now hunters can climb up into their raised hides at exactly this time, and they don't have long to wait until their prey appear.

In the final tally, however, it is the wild boar that seem to have come out ahead. In some cases, they have come to benefit from the bait as their main source of food and multiply quickly despite the hunting pressure – so much so that reducing populations has become a lost cause in many places. In the US, feral hogs are generally diurnal, if left undisturbed, but they react like their European counterparts, and intensive human activity or hunting during the day can drive them to be active at night.

There have been many particularly touching discoveries made about domestic pigs, if only because a number of research facilities are working to improve factory farming. When the newspaper *Die Welt* asked Professor Johannes Baumgartner at the University of Veterinary Medicine in Vienna if there had been any notable characters among the pigs he had studied, he told them about one old sow. Over the course of her life, she gave birth to 160 piglets. She taught all of them how to build a nest out of straw and, when her daughters grew up, the old sow assumed the role of midwife and helped them prepare for the births of their own babies.[21]

The question then becomes: if researchers know so much about the intelligence of pigs, why isn't the image of the smart pig publicised more? I suspect it has to do with eating pork. If people knew what kind of an animal they had on their plate, many would completely lose their appetite. We already know this from primates: could any of us eat an ape?

Gratitude

It should be clear by now that whether they are driven by their circumstances or our desires, whether they want to or not, animals love people (and, of course, the reverse is true). I consider gratitude to be a closely related emotion. And animals can certainly feel gratitude, as well. Owners of dogs with chequered pasts that have been welcomed into families later in life are particularly well placed to confirm this.

Our cocker spaniel, Barry, didn't come to us until he was nine years old. Actually, after the death of our Münsterländer, Maxi, we wanted to draw the dog chapter of our lives to a close. Or so we thought. Despite the fact that my wife, Miriam, was absolutely opposed to a new family member, our daughter set out to convince us otherwise. She didn't get much resistance from me, because I couldn't really imagine life without a dog. When my daughter accompanied me to an autumn market at a nearby country shop, both of us knew what was going to happen. The Euskirchen animal shelter was going to have a parade of its guests, hoping to find homes for them that day. My daughter and I were hugely disappointed when the only animals on display were rabbits, because we already had plenty of those at home. After waiting around the market all day, making multiple tours of the stalls, here we were, confronted with the fact that there were no dogs. Right at the very end there was an announcement that a future occupant was going to be shown by one of its former owners, before being delivered to the shelter: Barry. Our hearts beat faster. The dog was apparently extremely good-natured, a model passenger in the car and he was neutered. Perfect! We leapt up from the bench and stepped forward. A short test-walk, a hand-shake to seal the agreement for three days' probation and we took off right away with the dog in the car, headed for Hümmel.

The three-day trial was important, because Miriam didn't suspect anything yet. She came back late that night after an engagement. She was taking her coat off when my daughter asked: 'Do you notice anything different?' My wife looked around and shook her head. 'Then take a look down at your feet,' I prompted. And in that instant, it happened. Barry looked up at her, wagging his tail, and my wife took him into her heart right then and there, for the rest of his life. And the dog was grateful – grateful that his long and arduous journey had finally ended. His owner, an old lady suffering from dementia, had had to give him up. He'd gone through two different families, and now he had found his 'forever home' with us. It's true that for the rest of his life he worried that there might be yet another handover, but other than that Barry was always happy and friendly. He was grateful. It was as simple as that – or was it?

After all, how are you supposed to measure gratitude or – almost as difficult – to define it? If you check on the Web, you'll find a lot of discussion, but nothing definitive. Some animal lovers think of gratitude as their due, a response that many owners expect from their animals in return for the care they give them. I wouldn't even bother to search for this kind of gratitude in animals, for it would merely be an expression of subservience, smacking of servility. Essentially, and this is in reference to people, what emerges from most definitions is that gratitude is a positive emotion arising from an enjoyable experience caused by someone, or something, else. In order to be grateful, you need to be able to recognise that someone (or life) has done you a good turn.

The Roman politician and philosopher Cicero considered gratitude to be the greatest of all virtues, and he thought dogs were capable of feeling it. But now it gets tricky. How can I know whether an animal recognises who or what has caused its enjoyable experience? In contrast to the joy itself (which is easy to recognise in a dog), there's also the question of whether the dog gives any thought to the cause of its joy. It's relatively simple to answer this question. There's food, for starters. The dog is happy about its meal and knows exactly who filled its bowl. In fact, dogs often encourage their owner to repeat the process. But is this really gratitude? You could just as easily call it begging. Doesn't true gratitude include a mindset, a way of looking at life? An ability to celebrate small pleasures, without constantly craving more? Seen from this perspective, gratitude is when joy and contentment about circumstances that are not of your own making coincide. Unfortunately, this kind of gratitude cannot yet be proven in animals – we can do no more than speculate about their inner outlook on life. In Barry's case, at least, my family and I are certain that he was both happy and content to have found his final home with us, even if we don't have any scientific proof.

But how about other examples in the animal world? Might they shed more light on the issue? There is the story of a humpback whale in the Sea of Cortez off Mexico that put on an hour-long display of breaching and flipper-flapping, after a man called Michael Fishbach spent hours cutting off a fishing net in which it had been hopelessly entangled. When Fishbach encountered the whale, it looked as though it would not be able to survive for much longer. Fishbach immediately entered the water, armed with just a small knife. As soon as the whale was free, it put on a glorious acrobatic display. Perhaps it was just happy not to be entangled in the net any longer – or perhaps it was performing for the people in the small boat to thank them for rescuing it from certain death.[22]

Then there is this story of gratitude closer to home about wild birds that voluntarily shared their treasures with a young girl in Seattle. Crows like to collect shiny things and, of course, crows also like to eat. When Gabi was four, like many four-year-olds she was not a particularly neat eater, and sometimes she dropped her food on the ground when she was outside. The crows lost no time snatching it up. A couple of years later Gabi began intentionally sharing her lunch with the crows as she walked to the school bus stop. Then she began feeding the crows on a daily basis in her back garden. Soon after that, the birds began bringing her gifts: bits of glass, screws, pieces of bone, broken jewellery. One time Gabi's mother, Lisa, dropped the lens of her camera when she was photographing the crows and thought it was lost, only to find it right there a few days later on the bird feeder. The crows had even carefully washed it before returning it. Is this gratitude, perhaps? Crows are certainly known for recognising people and for having strong emotional reactions to those they don't like. In this case, it seems they liked Gabi very much, and perhaps they were grateful and were thanking her for looking after them for all those years.[23]

Lies and Deception

CAN ANIMALS LIE? If you define the term loosely, then quite a few can. The hoverfly, whose yellow and black stripes make it look like a wasp, 'lies' to its enemies by making them believe it is dangerous. It must be said that the fly is unaware of its deception, because it doesn't actively undertake it; it was just born looking that way. It's the same with the European peacock butterfly. With big 'eyes' on its wings, it signals to its enemies that it's bigger than it really is and too large for them to tackle. But let's put these examples of passive lying aside and take a look at which animals are the real tricksters with deception on their minds.

One of these, for instance, would be our rooster, Fridolin. He's a portly representative of his species and as white as driven snow, which is just as he should be because he's a white Australorp. Fridolin lives with two hens in a nearly 200-square-metre run designed to keep out foxes and hawks. Two hens are quite enough to provide us with the eggs that we need. Fridolin, however, sees things differently. Such a small flock doesn't keep him as busy as he'd like to be and, with a sex drive like his, he could easily satisfy a couple of dozen lovers. Circumstances, however, dictate that he has to concentrate all his loving on Lotta and Polly. The hens are not at all keen on constant coupling and therefore quickly give Fridolin the slip as he prepares to make that final pounce. If he somehow manages to land on the back of one of his ladies, despite her best efforts to avoid him, he spreads his wings to keep his balance. At the same time he grabs hold of the neck feathers of the hen, which is now squashed down flat on the ground, and sometimes he even pulls feathers out, in a frenzy of passion. Then he presses his cloaca against hers and squirts his semen inside. As soon as the act is over – it lasts just a matter of seconds – the hen gets up and gives herself a shake and, at least for the time being, she can return to eating unmolested. But it's not long before Fridolin's ardour returns, and because neither hen is interested in joining in, he has to chase them down all over again. The rooster often runs out of steam while pursuing them, and a modicum of peace descends.

But then Fridolin came up with an easier way of getting what he wanted. Fridolin is usually a real gentleman, allowing his little harem first dibs at food. As soon as he spies something tasty, he makes a special clucking sound, and it's not long before Lotta and Polly rush over and fall upon the food he has found for them. But sometimes there isn't anything to eat under Fridolin's feet, and it turns out that the rooster has enticed his hens over with a bald-faced lie. Instead of tasty worms or scrunchy seeds, what awaits them is another of Fridolin's attempts to mate and, thanks to the element of surprise, his devious plan is often rewarded with success. However, if he tries this trick too often – and with two hens, a couple of lies suffice – then both of them become cautious, even when real food is on offer. No one believes a liar, even when he tells the truth . . .

Other species of bird can also be big fibbers. Take swallows, for example. If a male returns to find his mate is not on the nest, he gives an alarm call. His mate wrongly supposes danger threatens, and she flies back to the nest via the shortest route. The male uses the false alarm call to keep his mate from pursuing dalliances in his absence. Once the eggs have been laid, this is no longer a concern, and the deceitful calls cease.[24]

Great tits provide another example. They are widespread in many parts of the world and a number of them are fibbers, for, when it comes to food, it's every bird for itself. These pretty birds with black heads and white cheeks have a sophisticated language that they use to warn each other about predators. One of these predators is a small raptor called a sparrowhawk, which likes to hunt in back gardens. The bird dives down, quick as an arrow, to grab sparrows, robins and tits, and then flies to the nearest bush to eat them. A great tit that spies the danger from afar warns others of its kind with a high-pitched call. The call is out of the sparrowhawk's range of hearing, which gives the great tits a chance to slip away safely before the raptor spies them. If, however, the raptor is already dangerously close, the warning is broadcast at lower frequencies, so that all the tits know an attack is imminent. The attacker can also hear this deep churring call and immediately realises that its planned surprise attack isn't a surprise any more. It often ends up grabbing thin air because the tits are on the alert. I'm sorry to report that some great tits take advantage of their well-functioning community. If there's particularly tasty food around or if food is scarce, then the little liars make the usual alarm call. All the birds quickly fly to safety – or almost all. Left alone, the trickster can now eat as much as it likes.

How about cheating on your mate? This form of sexual liaison is also a kind of deception – at least it is when the cheater knows what it is doing. And you can see exactly this with male magpies. Some urban communities consider these handsome black-and-white corvids to be public enemy number one, because they snatch the babies of other songbirds to feed their own young. This behaviour puts them in the same league as the squirrels I've already talked about. I like to imagine what it would be like if magpies were an endangered species. If that were the case, we would be excited to see them, and we would marvel at how their black feathers shimmer blue-green in the light. Unfortunately, people who appreciate their beauty are few and far between.

But back to cheating. Like other corvids, magpies form pair-bonds that last a lifetime. They set themselves up with their partners in a territory that, like their partnership, provides their home for many years. They vigorously defend their home from others of their kind. This is clearly because both partners want to avoid sexual dalliances; after the eggs have been laid and most of the business of procreation has been accomplished, the zeal with which they defend the boundaries of their territory diminishes considerably. But even before that happens, much of this territorial defence is strictly for show, at least on the part of the males. While the female aggressively drives off any competitors that intrude, her partner is an opportunist. As long as his mate is watching or within earshot, he'll be just as aggressive in driving off any intruding female. However, if he thinks he's not being observed, he begins eagerly courting the attractive stranger instead.[25]

There are, however, other strategies in the animal world that cannot really be called lies, even if this is how they are sometimes described in the media. There are reports of foxes that, in contrast to peacock butterflies, actively fool others. The fox's hunting repertoire includes playing dead, and a fox may even let its tongue hang out to make the scene more convincing. A corpse out in the open? There are always takers, most often crows, which are happy to help themselves if there's delicious meat on offer, even when it's a bit past its prime. In the case of the fox, it's super-fresh – too fresh, as it turns out. Any black-feathered guest that decides to grab a bite suddenly finds itself in the jaws of the wily fox and ends up being the meal instead.[26] That is a masterful piece of play-acting and definitely a trap, but it is far from being a lie. Lying usually involves deceiving others of your own kind by providing them with false information in order to gain an advantage for yourself. The fox is simply following a hunting strategy that is cunning and morally above reproach. The fox is quite different from Fridolin or from the magpie that sneaks a bit on the side, both of which are intentionally pulling a fast one on fellow animals near and dear to them.

No matter how you might judge these cases of trickery from a moral standpoint, I, for one, am moved by them because they show just how intricate the inner lives of animals really are.

Stop, Thief!

IF LYING IS WIDESPREAD AMONG ANIMALS, then what about theft? A good place to start is by looking at socially oriented animals, because thievery, like lying, has a moral component, and it is judged to be negative only when it is socially relevant and when it adversely impacts on other animals of the same species.

When it comes to thievery, the American grey squirrel is a sly one, but before delving into its behaviour, let's check out how it is doing in Europe, where it has become a serious threat to native red squirrels (which can also be black in colour). In 1876 a certain Mr Brocklehurst from Cheshire took pity on a pair of American greys that were being held captive, and he released them. In the years that followed, many dozens of other animal lovers followed his example. The grey squirrels thanked their liberators by reproducing enthusiastically – so enthusiastically that they have now pushed their red European relatives almost to the brink of extinction. Grey squirrels are larger and more robust than their red cousins, and they adapt to any kind of woodland, whether deciduous or coniferous. Even more dangerous for native European squirrels, however, is a stowaway that hitched a ride along with the greys: squirrel pox virus. Whereas American greys are generally immune to the virus, European red squirrels die in large numbers when they are infected. Unfortunately, greys were not only released in the UK; they were also released in northern Italy in 1948, and since then they have been advancing on the Alps. We don't know if at some point they'll manage to scale the mountains and march victorious into the German woods as well.

Despite all this, I don't want to characterise grey squirrels as vermin. After all, it's not their fault they were brought to Europe, and their dominance is not due to their behaviour, which brings us back to the topic of thievery. Squirrels sometimes get food by plundering the winter caches of other squirrels. In many cases, they must do this to survive, as the unsuccessful searches in the snow that I observe from my office window every winter attest. A squirrel that cannot remember the location of its caches will starve and, as a last resort, it may help itself to its neighbours' loot. I don't know whether European red squirrels have developed a counter-strategy, but researchers discovered that American greys have. A team from Wilkes University in Philadelphia observed grey squirrels digging fake, empty caches. They did this in clear sight of other squirrels to lead them astray; in fact, they only did so when they thought they were being watched. When the eyes of other squir-rels were on them, they dug around a bit in the dirt and pretended to bury something. According to the researchers, this is the first evidence of rodents using deceptive tactics. When many other squirrels were watching, up to 20 per cent of the caches were empty. As part of the experiment, the researchers then had students raid the caches that had food in them, and guess what happened? The squirrels reacted immediately and from then on used deceptive tactics in the presence of suspect humans, as well.

Theft is also a big deal for Eurasian jays. These birds are obsessed with securing enough food for themselves. In the autumn each bird caches up to 11,000 acorns or beechnuts in the soft forest floor, even though they could survive the winter just fine with far less food. They not only rely on the oil-rich seeds as emergency rations until the next growing season, but also feed them to their chicks in spring. Even allowing for this, the sly birds usually store way too many seeds. And what an amazing memory they have: jays find every one of their thousands of caches with a single stab of their beak. Small trees sprout from the unused seeds to ensure that future generations will have their own supply of nuts and acorns.

In the woodland I manage, we use the birds' passion for collecting to plant young deciduous trees in the monocultures of old spruce plantations. This is how it works. We put seed trays on posts and fill them with acorns and beechnuts. Jays love to come and help themselves, and they then distribute their booty in the soil hundreds of metres in every direction. It's a win-win situation. We get precious new stands of deciduous trees in the woodland, and the jays get huge quantities of winter provisions with very little effort. Some years, however, the oaks and beeches do not set seed, and then things get tight for these colourful birds. Whereas the population increases in years of plenty, in lean years it shrinks. This ruthless natural cycle has been repeated over and over since time immemorial. But who wants to starve? Some of the birds fly south, but most try to survive in the woodlands they call home.

Just like the squirrels, in lean times jays watch other jays in late autumn to see where they bury their treasures. And because no bird can keep watch over such a large number of hiding places, sneaky individuals can live well over the winter by profiting from the hard work of others. Scientists at the University of Cambridge have discovered that the birds are well aware of these shenanigans. They discovered this by putting trays filled with two different materials into the aviaries. Some trays contained sand, others gravel. Whereas sand doesn't make any noise when you dig in it, gravel gives the game away by rattling. And the jays kept this in mind when they buried their caches.

If the jays were alone in their enclosures, it didn't matter to them whether they hid the proffered peanuts in sand or in gravel. If the competition could see and hear them when they were digging, it also didn't matter which material they were rummaging around in. In the first case, no other bird was around to witness where the precious booty was hidden. In the second, the birds realised that any bird watching them would know where the food was anyway. However, if the competition was out of sight but still within earshot,

the jays opted for the quieter sand. Under those conditions, there was a much higher likelihood that the
potential thief would have no idea anything had been
hidden. And for their part, the thieves were also quieter.
Whereas they normally called loudly when they saw
other jays, when they were watching food being hidden
they were considerably less vocal, clearly to avoid
betraying their presence.[27] This experiment clarified
two things. First, the bird doing the hiding can put
itself in the position of other jays and take into account
what they can and cannot see. And second, the future
thief was obviously planning its actions in advance,
because it limited the sounds it made in order to
increase its chances of later plundering the cache of
peanuts undisturbed.

Of course, theft in the sense of the intentional
seizure of assets that don't belong to you doesn't
occur only within a species. Come winter, in many
deciduous woods you can find traces of interspecies
plundering. Sometimes you come across holes in the
forest floor 30–60 centimetres deep, their edges
strewn with big clumps of excavated earth. Wild boar
are the only animals that root around like this, and
they do it in so-called 'mast years'. This technical
term describes years when beeches and oaks go into
overdrive to produce seeds. These years were once a
blessing for farmers: a mast year meant they could
drive their domestic pigs into the woods one last time
to fatten them up before the winter slaughter. Farmers
are not allowed to pasture their animals in woodlands

any longer, at least not in Central Europe, but the term 'mast' – from the German *mästen*, 'to fatten' – is still used.

Naturally, wild boar behave the same way as their domestic relatives in a mast year: they put on a nice thick layer of fat. But once the unexpected gift has been cleaned up and all the nuts lying on the ground have been scarfed down, rumbling stomachs demand a top-up. And snacks can still be found deep down in the soil. This is where mice have buried stores – pantries stocked with their portion of the harvest, so they can make it safely through the winter. Even in times of hard frost the ground doesn't freeze more than a few centimetres below the insulating layer of leaf litter, and in the mouse's quarters it's always at least 5 degrees Celsius. Thanks to a cosy layer of leaves and moss and a completely draught-free location, a mouse can survive very well down here. That is, as long as no wild boar come calling.

The grey hustlers have extremely sensitive noses and can smell the mouse's digs from metres away. From experience, they know that the little creatures diligently collect beechnuts and other seeds, and store them all together in one convenient spot. An enormous store that will last a mouse for months is a tiny snack that will tide a wild boar over until its next meal. However, since mice mostly live in large colonies, a number of small snacks can provide the calories a wild boar needs to make it through a cold winter's day. And so the boar burrow along the underground tunnels, smashing storerooms and emptying them in a couple of gulps. The only option the mice have is to flee and face an uncertain future, for in winter there are very few sources of food for the homeless. If the mice can't evade the wild boar underground, they are gobbled up along with their food stores; boar enjoy their meat with veggies on the side. At least the swallowed mice are spared a long, slow death from starvation.

And what does this behaviour look like from an ethical point of view? The wild boar's plundering of mouse assets isn't really theft, because they are not deceiving other boar. They are perfectly aware that they are raiding mouse provisions, but this is a completely normal way for them to get food, even if the proceedings look very different from the perspective of the mice.

Take Courage!

IF ANIMALS FUNCTIONED ONLY according to fixed genetic programming, then each individual of one species would react the same way under the same circumstances. A certain amount of a hormone would be released that would trigger the corresponding instinctive behaviour. But that is not the case, as you probably already know from observing domestic animals. There are courageous and anxious dogs, aggressive and super-gentle cats, jumpy and bombproof horses. The character each animal develops depends on its individual genetic predis-position and, just as importantly, on the influence of its environment, which is to say, its life experience.

Our dog Barry was a little wimp. As I have mentioned, before he came to us, he had already been passed along by a number of different owners. For the rest of his life he was scared of being abandoned, and he always got extremely worked up when he was taken along while we visited friends. If you are a dog, how are you supposed to know whether you're going to be handed away yet again? He showed his nervousness by panting non-stop, so we finally gave up, leaving the distressed animal alone in the house for a couple of hours instead. When we got back, it was easy to see that Barry was relaxed. He became deaf in his old age and couldn't hear us arrive, sleeping soundly until he blinked up at us when he felt the wooden floorboards vibrate under our feet. So Barry is an example of an animal that lacks courage, but we want to take a look at the opposite trait, and to do that, let's step out into the woods.

One fawn that had breached a plantation fence along with its mother showed particular courage. I used to erect these fences around areas where storms had toppled trees in monocultures of plantation spruce. In order to allow as natural a woodland as possible to regenerate, forestry workers planted little deciduous trees. These newly planted areas needed to be protected from the greedy mouths of browsers, and that's why I erected the fences. The wire fences behind which the oak and beech saplings grew were 2 metres tall. During a late-season storm a spruce had fallen on one of these fences, flattening it. Deer, including the aforementioned doe and her fawn, had wandered through the gap directly into a land of ease and plenty. No walkers disturbed them there, and they could munch away on the tasty shoots of much-sought-after deciduous trees. Things looked a little different to me. The expensive fence was no longer of any use, and the goal to have a halfway natural beech and oak wood one day was fast disappearing. And so, accompanied by my Münsterländer, Maxi, I climbed in after the deer to drive the freeloaders back out.

To do this, I made an opening in one corner of the fence so that when I drove the deer along the inside of the wire, they could escape. Once Maxi got in on the action, this was the only way they could go. She responded to my hand signals from about 100 metres away, dashing here and there, flushing the intruders out of the shrubby undergrowth. One of the deer that had found its way in rushed past me and out of the opening, but then, 20 metres further along, crawled back in on its stomach through a tiny gap under the fence. Things weren't working out with the doe and her fawn, either, on account of the fawn. The mother was now trying to lead it out at a gallop, and Maxi was chasing it at full tilt in the direction it needed to go, when suddenly it all got too much for the youngster. It turned round and rushed menacingly towards the dog. Maxi was normally very courageous and there wasn't much that frightened her, but a fawn coming right at her? She had never experienced anything like it. She pulled up, baffled, but the fawn continued its charge. Maxi finally turned and fled. That was it for me that day, and the deer were allowed to stay in the newly planted area. They had completely lost respect for the dog, and all I could do was shake my head and laugh. I'd never encountered such courage in a young fawn. And it had shown courage, because its mother had failed to step in and defend her offspring.

But what exactly is courage? Once again, this term has a variety of vague definitions – I invite you to try to come up with one off the top of your head – although one general concept seems clear: courage involves realising that it is important to act despite recognised danger, and then doing so. In contrast to bravado, courage is considered to be a positive quality and, in this sense of the word, the fawn had certainly acquitted itself well.

Just as courageous, by the way, are the fieldfares I mentioned earlier that raise their broods in the old pines by our forest lodge. When a carrion crow, their old arch-enemy, shows up, they don't just sit there and watch it grab their chicks. As soon as the feared bird starts to fly towards the colony, an aerial attack is mounted. The little thrushes swarm around the considerably larger intruder and dive-bomb it incessantly. It would be easy for the crow to fight off or injure the tiny raging birds, but their determined pre-emptive strikes, usually mounted in concert with other thrushes, throw the crow off and it starts weaving to evade its attackers. It doesn't notice that the attacks are driving it away from the nest, which, of course, is what the thrushes want. It seems to find the attacks really annoying, because after a few minutes the crow usually beats a retreat and disappears from the stand of old trees. Does that mean that field-fares are courageous? Or are they simply running a genetic programme activated by the appearance of their enemy? It's a mixture of both, and that's just how it is in every similar situation – probably even with us. Not all thrushes react in such a spirited fashion or, above all, with such persistence. How far they pursue the crow, and how fiercely they dive-bomb it, is different with every bird. And while fearful thrushes mount only half-hearted attacks, the courageous masterfully manage to drive the crow off for several hundred metres.

But are the less brave birds automatically at a disadvantage? Niels Dingemanse and his team at the Max
Planck Institute for Ornithology don't think so. They
investigated great tits for similar character traits and
discovered that the shyer individuals get on better with
other great tits. They avoid disputes and large flocks
and prefer to live in small groups of like-minded birds.
Shy birds are slower and calmer, and take their time
before mobilising. And this means they notice things
that their courageous, speedy colleagues often overlook,
such as seeds left over from the previous summer.[28]
Because the advantages and disadvantages to being
either courageous or shy seem to balance out, both
character traits have survived until today.

Black and White

MANY PEOPLE ARE INTERESTED in animal emotions in principle, but their interest doesn't usually include all species, particularly not those considered dangerous or disgusting. People often ask me: 'What use are ticks?' The question amazes me to this day, because I do not believe that any one animal has a more important job to do in an ecosystem than another. That might sound odd coming from a forester, but I think a principle like this extends to all creatures the respect they deserve.

But let's take this one step at a time. First, let's look at more insect examples – wasps, for instance. In late summer these social insects can get really annoying, and at some point even I've had enough of these stripy little stingers. Perhaps this goes back to an experience I had in my youth. I was racing along on my bicycle on the way to the swimming pool, when a wasp flew at me and, because the wind was in my face, it got stuck between my lips. I pressed my mouth shut, but I could do nothing to stop the wasp from stinging me liberally. It felt as though my mouth was being sewn shut. My lower lip immediately swelled up so much I was afraid it was going to burst. It didn't help that at that age you tend to be very self-conscious when it comes to physical blemishes. Since then, I've not been very keen on wasps.

You might have had a similar experience yourself, so it's hardly surprising that you can buy all kinds of gadgets to keep wasps away. There are those bell-shaped glass contraptions that you fill with enticing sugary liquids to attract the wasps and then drown them. It sounds mean, and it is. But stinging insects are usually not held in high regard, and we're happy to dispatch them quickly.

Let's move on to a row of cabbages growing in a raised bed belonging to a colleague of mine. A whole bunch of fat cabbage-white caterpillars were sitting on the succulent leaves. Gardeners categorise these caterpillars as pests, because they riddle cabbage leaves with holes all the way down to the ribs. My colleague asked us for advice and we were able to help her out: we had had good experiences with neem oil for a number of years. Once we started using this ecologically friendly spray (which is also certified for use in organic agriculture), our cabbages made it through to harvest intact. But we're not using neem oil any more, and this is where the wasps come in. They pounce on the caterpillars and munch them into little pieces, so they can carry their bite-sized booty back to their nests to feed their hungry brood. And just like that, all the caterpillars disappear, as we can attest from our experience at our forest lodge. A plague of wasps in the summer means caterpillar-free cabbages. Does that mean wasps are beneficial?

Most of the creatures in our garden have earned labels like this. Tits: beneficial (eat caterpillars). Hedgehogs: beneficial (eat snails). Snails: pests (eat

greens). Aphids: pests (suck the juices out of plants).
How wonderful that for every creature that is considered
a pest there is a beneficial one to keep it in check. But
if you divide Nature like this, you automatically assume
two things. First, there must be a Creator who has
planned and implemented a design where everything is
finely tuned and in balance. Second, this Creator has
made our world in such a way that it is completely
oriented to our needs. Questioning the point of ticks
makes sense in this worldview. I don't want to criticise.
After all, this way of thinking is widespread even among
conservation groups that encourage beneficial creatures
by, for instance, building nest boxes for them. But can
Nature really be neatly compartmentalised like this?
And if it can, where would we fit?

No, I believe that the immensely vibrant lives of
millions of species have adapted so well to one another
only because overly selfish species end up recklessly
exploiting all available resources, destabilising ecosystems and irrevocably changing them and their inhabitants. Just such an event played out about 2.5 billion
years ago. At that time many species lived anaerobically,
which is to say without oxygen. For life at that time,
the most important gas we breathe today was pure
poison. But then cyanobacteria began to spread with
astonishing speed. They got their food by photosynthe-
sising and, as they did so, released a waste product into
the air: oxygen. At first this gas was taken up by rocks,
and rocks that contained iron, for example, rusted. But
at some point there was so much excess oxygen that

the air became increasingly oxygen-rich until, finally, a deadly threshold was crossed. Many species died out, and the ones that remained learned to live with oxygen. At the end of the day, we are the descendants of the creatures that adapted.

Basically, small adaptations are being made every day. What we understand as a finely tuned balance between prey and predators is in reality a harsh struggle, with many losers. When a lynx is wandering through its enormous territory, it's hungry for deer. However, the cat is not a good sprinter and so it must rely on the element of surprise. Clueless, careless browsers that have not yet heard the news about a large predatory cat in the area are particularly easy prey. A lynx can enjoy a deer a week, but only until all the deer in the area have been tipped off to its presence. Then panic prevails at the snap of a twig, and even pets are wary. A colleague of mine tells me that his cat is the first to alert him when there's a lynx in the area. The cat refuses to set foot outside, but he can't say what alerts the cat. Perhaps it is the collective behaviour of potential prey that gives rise to an aura of suspicion in the wood. Because of this, the lynx makes fewer kills and is forced to move on. And it is only a few kilometres farther on, in a new area of unsuspecting prey, that it can get back to easy pickings. If there are too many lynx on the prowl in the same area, at some point there will be no more unsuspecting prey. Then, when winter ushers in lower temperatures and a corresponding rise in energy needs, many lynx starve – especially the younger, less experienced animals. You could say that the population is self-regulating, but what that really means is that animals die ghastly deaths.

So Nature is nothing like a neat set of compartments. No species are inherently good or bad, as we have already seen in the case of squirrels. But it is much easier for us to empathise with, or at least take an interest in, squirrels than it is for us to relate to the ticks mentioned at the beginning of this chapter. And yet even these abhorrent little creatures have feelings. The presence of something as simple as hunger is empirical proof of this, for the tiny arachnids go after mammal blood only when their stomachs are rumbling. And this means that an empty stomach must feel unpleasant, particularly when it hasn't been filled in almost a year, which is how long ticks can hold out until their next meal, if they absolutely have to.

When a large animal crashes by, ticks can feel the disturbance and smell sweat and other animal odours. They quickly extend their forelegs and, with any luck, grab onto the legs or body brushing by and hitch a ride. Eventually the ticks crawl to a comfortable, warm spot where the skin is not too thick and there they settle in. They push their extended mouthparts into the wound and slurp up the blood that seeps out. The tiny vampires can increase their weight many times when feeding, and they swell up to the size of little peas. On the way to maturity, they must undergo three moults, and before each stage they need to find a new victim to fill up on. That's why it can take up to two years for ticks to reach adulthood. When it gets to the stage when the smaller males and the larger females have sucked themselves full to bursting, all that remains is the finale.

The little males must mate. Must? They want to. Just like us, they are driven by desires, and they eagerly seek a partner they can grab hold of, to complete the final act. After that – and here, I hope, is where the similarities end – they die. The females live long enough to lay up to 2,000 eggs and then they, too, pass away. The tick is a creature whose ultimate joy, or – because this cannot yet be proven – at the very least the high point of its life, consists in producing thousands of offspring and then dying, completely spent. If it were a mammal, we would call this behaviour self-sacrifice. For ticks, unfortunately, we currently reserve just one emotion: disgust.

Cold Hedgehogs, Warm Honey Bees

WE ALL LEARNED this in our biology lessons in school. Apart from all the other distinctions you can make, the animal world can be divided into warm-blooded and cold-blooded creatures. Yet more compartments and, as you will see, animals don't fit neatly into these, either. But first let's get to the scientific categories. Warm-blooded animals regulate their own body temperatures and keep them constant. Humans are a good example. When we get cold, we shiver to generate the heat we need. When we overheat, we sweat and cool down as our sweat evaporates. In contrast, for better or worse, cold-blooded animals are dependent on external temperatures. When it gets too cold, they can forget about physical exertion. And that's why, every winter, I find flies in my woodpile that just don't have the energy to get up and go. When temperatures drop below freezing, they are not up for much more than crawling around excruciatingly slowly among the logs. Helpless, the most they can hope for is that no birds track them down when it's cold. And that's how it is for all insects. All of them? Actually no, that's not how it is for my bees (or any other bees, for that matter).

I used not to like bees. It's difficult to forge a relationship with insects, and if those insects sting, the almost automatic response is aversion. Apart from that, I rarely eat honey. Not an auspicious start for a bee-keeper, although that's what I've become. It's really only because of the apples. We had hardly any bees in our orchards in spring. To change that, I acquired two hives in 2011. Since then, pollination has not been a problem, and we've had plenty of honey. But, most of all, I've learned that bees are different from other insects. For example, in some respects they are similar to warm-blooded animals, and that is the main reason they are so eager to forage. Nectar, processed into honey and stored in combs, serves as a winter fuel for bees, for they love to be cosy and warm. Their comfort zone lies between about 33 and 36 degrees Celsius, just a little cooler than the body temperature of mammals.

In summer, that's not a problem. Quite the opposite, in fact. The muscle action of up to 50,000 busy bees creates a decent amount of heat, which must be dissipated so that the hive doesn't get too hot. This is a complicated task. Worker bees carry in water from the closest source so that it can evaporate and cool the interior of the hive. Thousands of wing-beats circulate air in the hive, creating a cool breeze between the honeycombs. Too much disturbance overwhelms this communal effort. If the hive is attacked from the outside, or if it is handled incorrectly by the bee-keeper while being transported from one place to another, the agitated bees heat up so much that the combs melt and the insects die of heatstroke. The technical German term for this translates as 'death by buzzing'. The expression comes from the noise the colony makes as the bees panic and beat their wings faster, heating up the hive and causing their own demise.

In the normal course of events, however, the bees' thermoregulation processes work perfectly. For my hives, most of the year it's too cold rather than too hot, and creating heat is most important. Vibrating muscles means calories spent, and bees take in the energy they need in the form of honey. Honey is basically a thick, highly concentrated sugar solution with vitamins and enzymes added. Every month, primarily in winter, my bees burn through more than 3 kilos of honey per colony. Honey serves the same function for bees as fat does for overwintering bears, and just as bears emerge from hibernation much skinnier than they were in the autumn, so the size of bee colonies can shrink tremendously in cold weather. If it gets really cold, the insects huddle together and form a ball. It's warmest, and therefore safest, in the middle – and, of course, this is where the queen must be. But what about the bees on the outside? If the exterior temperature drops below 10 degrees Celsius, they would die of cold in just a few hours, so bees inside the ball are kind enough to take it in turns to give the outsiders the opportunity to warm up again in the dense, seething mass.

Bees prove that not all insects are exactly cold-blooded, and as you've probably guessed, not all mammals are exactly warm-blooded, either. Maintaining a constant body temperature is supposedly a mammalian (and avian) speciality. Supposedly. The little hedgehog is the exception that proves the rule. While the similarly sized squirrel bounces through the branches every once in a while during the cold season, even in the snow, this spiny ground-dweller spends its winters fast asleep. Its spines don't insulate as well as a squirrel's thick fur, and therefore it uses a great deal of energy when temperatures fall. Apart from that, its favourite foods – beetles and snails – have made themselves scarce and are no longer to be found above ground. So what better idea than to take a break as well? The prickly little guys roll up into a comfortable ball in a cosily padded nest that is often buried deep beneath a pile of leaves or brush. Here they fall into a deep sleep that can last for months. In contrast to many other mammals, instead of keeping their body temperature at a hedgehog-appropriate 35 degrees Celsius, they simply shut off their energy intake, which means that their body temperature falls to match the ambient temperature and sometimes drops as low as 5 degrees. Their heartbeat slows from up to 200 beats to only nine beats per minute, and they breathe only four times a minute instead of fifty. Dialled down like this, a hedgehog uses hardly any energy at all and can make it through to the next spring on its reserves.

Hedgehogs don't mind the cold – quite the opposite, in fact. The strategy I have just described works best when it is cold and frosty outside. It becomes dangerous only when winter temperatures rise above about 6 degrees Celsius. Then the hedgehogs slowly rouse, and the deep sleep of hibernation turns into a lighter sleep during which considerably more energy is used, but the

animals are not awake enough to move. If such conditions persist for a long time, some of the little sleepers
starve. It is only when temperatures reach about 12
degrees that the hedgehogs can really get moving and
nibble on something – that is, if they are able to find
anything, because their prey are still hiding out in their
winter quarters. With any luck, a number of early risers
in this predicament will be found by humans and fattened
up in hedgehog rescue centres.

And what do hibernating hedgehogs dream of?
During the really deep phase of sleep, metabolic rates
are extremely low and there's hardly any dreaming
going on, because the brain uses a lot of energy in the
highly active state of dreaming. Therefore without
metabolic activity, there are definitely no movies being
played in their minds. But what about that light sleep
above about 6 degrees Celsius? If hedgehogs can dream
then – after all, their energy use rises – the pictures
might be more like nightmares from which the hedgehog
would like to wake, but can't. Whatever is going on
inside its head, the situation is life-threatening and
perhaps the animal realises this in its drowsy state as
it struggles in vain to wake up. Poor little thing. Climate
change will, unfortunately, bring more of these warm
winter interludes.

Things are somewhat better for squirrels, at least
in the dream department. They don't hibernate, they
just doze for two or three days at a time before they
wake up again and feel hungry. Although their heart
rate drops during these down-times, so they use fewer

calories, their body temperature remains high. This means they need regular servings of energy-rich foods such as acorns and beechnuts. If there are none or if the squirrels cannot find them, they starve. And the strategy of red deer is much more like that of hedgehogs because, surprisingly enough, they too can lower their peripheral body temperature. They do this multiple times during the course of a day, meaning that their periods of winter 'hibernation' last only a few hours. Even such short down-times allow them to restrict their use of valuable body fat. Despite low exterior temperatures, their metabolic rate is then up to 60 per cent slower than in summer.[29]

Now another problem arises: red deer have to rev up their metabolic rate when they are digesting food, but going through the winter without eating anything is not really an option, either. When a red deer eats in winter, it often extracts less energy from the food than it uses to digest it. And that is why, paradoxically, when hunters feed red deer in winter, the animals can end up starving in droves. This is what happened in my home district of Ahrweiler in 2013, where there was an indignant outcry from hunters who wanted to continue feeding the animals despite a local ban. Almost 100 red deer died from starvation, a number of which would probably have survived if they hadn't been physically stressed after digesting the hay and sugar beets that hunters had fed them. Left to their own devices in winter, red deer mostly live off the body fat they accumulate in the autumn.

At some point I began to worry whether red deer constantly feel hungry in winter, which is a distressing thought. To stand in cold snow with a rumbling stomach and super-cooled extremities is surely very unpleasant – at least it would be for a human being. But it has now been proven that animals can turn off the sensation of hunger. Hunger is, after all, a signal from the unconscious that it's time to eat. And this feeling should only trigger the desire to eat when adding calories would be beneficial. Take hedgehogs, for instance. Even when they are hungry, they refuse smelly rotten food. The unconscious part of their brain temporarily shuts down their hunger pangs and replaces them with a firm resolve not to eat any of the proffered sustenance. We don't know whether red deer experience an aversion to buds and dry grass or whether they simply feel full. But we do know that the animals don't feel hungry in winter despite their fast, because at the end of the day not eating makes fewer demands on their energy reserves.

The combination of lowering body temperature and metabolic rate that I have just described doesn't work equally well for all red deer. How well it works depends on the character of the individual deer and, equally importantly, on its rank and position in the herd. Winter is particularly dangerous for red deer that have strong personalities. As leaders in the herd, they have to be constantly alert. This means their heart rate is continually elevated, and their energy use is correspondingly high. It's true that herd leaders have preferential access to good feeding grounds, but that is of little use to them.

Meagre winter offerings of dry grass and tree bark don't deliver sufficient calories to build up the fat reserves that these deer are burning through at a higher rate than their lower-ranking herd mates, which spend the cold winter nights standing around calmly and dozing. The lower-ranking deer are eating less than their leaders, but they are also using far less energy. And so, at the end of winter, they have more reserves than their superiors. After watching deer in large natural enclosures, foresters in Vienna made the surprising discovery that being the leader of a herd reduces a deer's chance of survival, despite the fact that leaders always get to help themselves to food first. According to the researchers, in future it will be more important to consider the life histories and personalities of individual animals rather than the average for the species. After all, that is exactly how evolution works – using deviations from the norm.[30]

Warm-blooded and cold-blooded, then, are fluid categories that can merge into one another. What about the sensation of freezing? This signals to the body that its temperature is falling dangerously low and measures must be taken to counteract this. For humans, it's lights out when our body temperature dips lower than 34 degrees Celsius. Before that, we start to shiver and we try to get somewhere warmer. It's the same for our horses. On cold, windy winter days our older horse, Zipy, is particularly prone to shivering, and she takes refuge in the pasture shelter. As the mare has less fat and muscle mass than her companion, and as her body is less well insulated despite her winter coat, that's sometimes not enough. On really cold days, we put a rug on her to warm her up until the shivering stops and she feels better. It's quite clear that feeling cold is as unpleasant for Zipy as it is for us.

But what is it like for insects? Their body temperature fluctuates along with the air temperature, and they don't have a mechanism to keep it at a certain level. In autumn, insects burrow into the ground or under tree bark or into plant stems so that they don't freeze solid. To ensure that any ice that forms in their cells doesn't cause the cells to burst, they store substances like glycerine, which inhibits the formation of larger and sharper ice crystals. What might that feel like? Do such creatures even experience cold? When I watch frogs and toads jumping into ice-cold ponds in late autumn in order to doze in the mud at the bottom, I can't imagine that they feel freezing cold. We find cold water so unpleasant only because it draws off our body heat far more efficiently than air. But if your body temperature is the same temperature as the water in the pond, it can't feel too bad to jump in. And so the frogs and toads down there are probably not feeling like they're freezing after all.

But if insects, lizards and snakes don't know what it's like to feel uncomfortably cold, does that also mean they don't know what it's like to feel pleasantly warm? I don't believe that. After all, in spring these creatures love to seek out sunny spots. The more their small body heats up, the more swiftly they can move. Therefore they perceive being warm as something positive, a condition that can be costly for some. Take slow-worms, for example. (Slow-worms are a kind of legless lizard, despite their common name.) Roads warm up particularly quickly in the sun. Tarmac stores heat and radiates it, even at night, so roads are a good place for slow-worms to fill up on warmth. Unless, that is, a car runs over the little sun-worshipper, which, unfortunately, happens quite often. Drama aside, it is clear that, like us, cold-blooded animals must also have a feeling for temperature, although they probably sense it differently than we do.

Crowd Intelligence

SOCIAL INSECTS BELIEVE IN DIVISION of labour. Scientists have coined the term 'superorganism' to describe a collective in which each individual is part of a greater whole. In the woods where I live, red wood ants with their enormous anthills are a typical example of this kind of arrangement. The largest anthill I ever found there measured 5 metres from side to side. There are usually multiple queens inside laying eggs to ensure the survival of the colony. The queens are cared for by up to a million workers, all of them female. The lowest on the social ladder are the winged males, which fly out to mate with other queens and then die. The female workers, with a lifespan of up to six years, are unusually long-lived for insects, but the queens, which live for up to twenty-five years, eclipse even this impressive achievement. Not in the literal sense of the word, I have to say, because ants need the warmth of the sun so that they can work. That's why they live in bright, airy coniferous forests.

In Central Europe, red wood ants have spread beyond their natural range, thanks to the proliferation of spruce and pine plantations. The fact that these ants are legally protected has less to do with their scarcity than with their reputation as 'wood police'. Supposedly, they help rid forests of troublesome pests such as bark beetles and caterpillars, although the red-black ants are blissfully unaware of their appointed role. And in addition to the aforementioned pests, they naturally also eat protected and extremely rare species. As ants, they have no concept of our habit of categorising creatures as harmful or beneficial, but that doesn't make their colonies any less fascinating.

Their relatives, the bees, live in a similar fashion and have been particularly well researched. Bees are also born into a society with a strict division of labour. There's the queen, which develops from a normal, fertilised larva. Whereas other baby bees are fed a mixture of nectar and pollen, larvae destined to hatch into royalty receive a special food called royal jelly. It's produced in the hypopharyngeal glands of worker bees, and while normal larvae develop into mature bees in less than twenty-one days, the turbo-diet of jelly produces a new queen after just sixteen days. A queen flies only once in her life: her nuptial flight. And it is during this flight that she mates with drones (male bees). After her return to the hive, for the rest of her life (about four or five years) she lays up to 2,000 eggs every day, interrupted only by short winter breaks.

For their part, the female worker bees spend every moment of their short lives hard at work. In the first days after they hatch, their job is to feed the larvae. After ten days they are also responsible for storing nectar and converting it to honey. Once three weeks have passed, they are finally allowed out into meadows and pastures to collect nectar for another three weeks.

Then they are worn out and die. Only the overwintering
bees that wait for the next spring, huddled together in
a tight cluster around the queen, get to live a little
longer. The sole task for the drones is to inseminate
queens, and because that happens only once, and only
a few of them get this opportunity, most of the time
they just hang around with nothing much to do.

Everything, then, is strictly pre-programmed, down
to the smallest processes. Inside the hive, the bees dance
to communicate information about nectar sources and
how far away they are. They turn nectar into honey by
adding glandular secretions and drying the mixture on
their tiny tongues. They sweat out wax and form it into
artful combs. Scientists recognise the bees' accomplish-
ments, but because, in their opinion, small insect brains
are unable to rise to intellectual heights, the individual
bees are considered to be components of a superorganism and their cognitive accomplishment is called
'crowd intelligence'. In such an organism, all the animals
are like cells working together in a much larger body.
Whereas the individual animals are considered to be
quite stupid, the interaction of the different processes,
as well as the ability of the whole to react to stimuli in
the world around it, is recognised as intelligent. This
perspective does not recognise bees as individuals;
instead, each bee is reduced to a building block, a piece
of a larger puzzle. Consequently, in the language of
German bee-keepers of old, the colony of bees was called
'the bee', referring to the population of bees as though
it were a single entity.

Whatever we think, it's all the same to the bees. But since I've been keeping bees, I now know this point of view is incorrect, because there's a lot more going on inside their little heads. For example, bees can definitely remember people. They will attack people who have annoyed them in the past, and allow people who have left them in peace to venture much closer. Professor Randolf Menzel at the Free University of Berlin has discovered other amazing things. Young bees leaving the hive for the first time use the sun as a kind of compass. With the sun as their guide, they develop an internal map of the landscape around their home and use it to record their flight paths. In other words, they have an idea about what things look like around them. In this way, they orient themselves much the same way we do, for people also create mental maps.[31]

But that's not all. The waggle-dance that returning worker bees perform for other bees indicates the abundance of, the direction of and the distance to a source of nectar, such as a field of rapeseed in full bloom. In his experiment, Randolf Menzel and his colleagues saw to it that the nectar source was then removed. When the frustrated bees returned, they got new coordinates by watching the dance of workers that had spotted blooms elsewhere. But the researchers then removed this second source as well, which meant more frustrated returnees. After this, Menzel noticed something completely unexpected. Some bees tried the first location a second time and, when they discovered there was still nothing there, they flew directly to the second location. But how did they manage that? All the waggle-dance had communicated was distance and direction from the hive. The only explanation is that the little creatures actively processed the information about the second location, so that they could find it from the first. You could say: they remembered, reflected and then rerouted. Crowd intelligence could be of minimal use to them here, and they must have relied on their own little brains to make this calculation. And there's more. In so far as they plan for the future, reflect on things they have not yet seen and are aware of their bodies in relation to these thoughts, bees are self-aware. 'The bee knows who it is,' says Randolf Menzel. And for that, it has no need of a crowd.[32]

Hidden Agendas

IF BEES KNOW WHO THEY are and plan for the future, what about birds and mammals? Whenever I'm animal-watching, I wonder whether the individuals I'm observing are consciously aware of what they're doing. That's very difficult for a layperson to determine — and that's what I am, despite my engagement with the topic. But as I explore the subject, I don't want to rely on studies alone; I also want to experience at first hand what animals are thinking. That might sound like asking for too much — it's difficult enough to work out what other people are thinking just by watching them — but during a conversation at the breakfast table, my children suggested that I had already experienced something along these lines, albeit only briefly.

I was telling them about the crow that waited for us every morning in the horse pasture. The black bird was always hanging around out there with a few other crows, and its territory must have been close by. Unfortunately, it's still legal to hunt and shoot crows where I live, and so these intelligent birds are very shy around people and normally keep a safe distance, usually about 100 metres. Over time, the crows in the pasture have become used to us, and they consider a distance of about 30 metres to be sufficient. That is, all except one, which is gradually getting tamer. On a good day, it allows us to approach within 5 metres, and every time we are touched by its trust. We talk to it, and it always gets a bit of grain, which we put on the hitching rail by the pasture gate. Aha – so it's all about food! The crow doesn't approach because it is curious or simply wants to be with us: no, it's well aware that when we appear, so does a meal. Even so, we enjoy seeing the crow on our daily rounds. We don't set the emotional bar very high, and that's just fine with us.

Because of this, on the morning in question I observed something that I found merely amusing at first. It was December. The pasture was soaked after weeks of rain, and every step I took in my heavy rubber boots sent up a spray of mud. It's hardly fun going out to feed the animals in weather like that, especially not when a side-wind is blowing a persistent drizzle into your face. Oh well. The horses were waiting for their morning ration of grain, and it's a well-known fact that moving around in fresh air always makes you feel better. To prevent the younger mare from immediately polishing off her older companion's portion, I had to wait, ready to interfere if Bridgi started helping herself to Zipy's food. Usually my presence was enough to ensure that the younger animal behaved herself, and in the minutes it took the horses to eat their breakfast I had time to watch the landscape – or, rather, the crow.

That morning, the black bird flew over from the nearby wood, having spied me a while before in my green-and-orange jacket with the white feed-bucket in

my hand. However, instead of flying to its usual observation spot, a post near the hitching rail, it landed in
the pasture only about 20 metres away. I suddenly
noticed it was carrying something in its beak: an acorn.
The crow then tried to hide its tasty treat. It poked a
hole in the ground, stuffed the acorn into it and pulled
a tuft of grass over the hole. I was marvelling at the
perfect camouflage when the crow turned in my direction. Had it noticed that I was watching? Immediately
it retrieved the acorn from its hiding place and began
to poke another hole in the ground. Just the one? No,
many holes, and at each one the crow pretended to stuff
the acorn inside. When it came to the last hole, the nut
disappeared and the bird was satisfied. After all, it had
gone to a great deal of trouble to deceive me and prevent
me from eating its favourite food. Only then did the
crow fly over and land on the rail, to eat the small
offering of grain I had left for it.

When I related this little story back at home over
breakfast, my children pointed out that it was a fine
example of forward planning. Then the penny dropped.
The whole time I had been amused at how the bird was
hiding its food from me, and that alone was highly
intelligent behaviour. After all, the bird had to consider
what I might have seen and how, when it was being
observed, it could hide the acorn in such a way that I
would be fooled. But right there in front of me, the crow
had considered something else, as well.

Even a crow's stomach can hold only so much, and
clearly an acorn meal would have filled it up. Of course

the crow could still have flown over to the offering of grain, but as it would already be full, its best option then would be to pick up and hide the grain. However, hiding individual kernels of grain is very labour-intensive, and so even though it was hungry, the bird first stashed the large acorn safely and then flew over to the rail to fill its stomach at leisure. Finally, it joined its companions in the neighbouring pasture, and I'm sure it retrieved the acorn later. That was perfect time-management to optimally exploit the offer of food, and to do that the crow must have given thoughtful consideration to the future. It told me to look more closely the next time I observed animals and, moreover, to think more carefully about what I was seeing. And who knows, perhaps you've had similar experiences and are now able to make sense of them with the benefit of hindsight.

Simple Sums

In my book, *The Hidden Life of Trees*, I reported that trees can count. In spring they make a note of the warm days above 20 degrees Celsius, and they leaf out only when a certain threshold has been exceeded. If these large plants can count, it makes sense to assume that animals have this ability as well. People have certainly wanted them to be able to count for a long time, and there have been many reports of amazing animals. Take Clever Hans, for example. This stallion could spell out words, read and count — or at least that is what his owner, Wilhelm von Osten, claimed when he made the horse a public attraction in Berlin early in the twentieth century. An investigative committee confirmed the horse's abilities, without being able to explain them. Eventually the puzzle was solved: Clever Hans was reacting to his owner's almost imperceptible facial expressions. The moment von Osten disappeared from the horse's sight, the horse's abilities vanished as well.[33] By the end of the twentieth century there was an increasing amount of hard data that confirmed that many animals can perform some rudimentary calculations. But mostly the data had to do with food and estimating larger or smaller quantities. I don't find it particularly interesting that animals can tell the difference. After all, when faced with the choice of more or less, deciding for more is part of what drives evolution, isn't it? What is much more interesting is whether real counting is possible.

Perhaps we can get closer to an answer by considering our goats. This time it wasn't me but my son who discovered what might be going on in the minds of Bärli, Flocke and Vito. While we were on holiday, Tobias took over responsibility for our little Noah's ark. Usually the goats get a small ration of grain at noon, and this is the highlight of their day. When it's time for their snack and we're in the pasture, the eager goats come running. In contrast, in the morning and evening, when we are 'only' feeding the horses, which are right there next to them in the pasture, the goats barely acknowledge our presence. Tobias changed the feeding times to suit his schedule, and every day was different. Sometimes the goats weren't fed until early evening and the horses' last feed was even later. When Tobias made his second visit to the pasture in the early evening, Bärli and her little family ran to him bleating, noisily demanding their portion of grain. It was, after all, my son's second appearance of the day, even though it did not coincide with the time of day when the goats were expecting a tasty treat. Is that evidence that goats can count? They always love grain, but they were now demanding it at what was, for them, an unusual time. Did they know this was Tobias' second visit to the pasture and so it was their turn to be fed? If it was nothing more than gluttony, then they would have begged, as so many domestic animals do, and demanded food every time a member of the family appeared. But they did so only on one out of three daily visits – the middle one.

Apart from our goats, is there evidence of this sort of intelligence in other creatures? It's pretty much old news by now that ravens are in the same league as apes, so let's take a look at pigeons instead. These birds have become a plague in our cities, and I must admit that it's not very nice to have a dollop of bird-poop land on your new jacket while you're standing on the pavement, which is what happened to me only recently. Despite the mess, the birds don't deserve the popular pejorative 'rats with wings': it is thanks to their intelligence that pigeons can establish themselves so persistently in our urban pedestrian zones. Professor Onur Güntürkün of the University of the Ruhr-Bochum has amazing tales to tell. His colleagues trained pigeons to recognise cards with abstract patterns. After training, the birds could differentiate between a staggering 725 different images. They were divided into 'good' and 'bad' cards that were presented in pairs. If a pigeon pecked at the good card, it got food. If it sunk its beak into the bad one, not only did it not get any food, but it also found itself in the dark (which pigeons can't stand). To pass the test, all the birds had to do was memorise just the good cards; however, the scientists were able to prove that the birds were not taking the easy way out, but actually memorised all the pictures, both good and bad.[34]

Our dog Maxi offered a completely different example of the ability to count, and it had to do with her ability to tell the time. She had the habit of sleeping soundly through the night until just before 6.30 a.m., when she would begin to whine softly, asking me to take her walkies. Why 6.30 a.m.? This was the usual time our alarm went off and the whole family got up to eat breakfast and then leave for school or work. It seems we could have managed fine without an alarm of our own as Maxi had a well-functioning internal clock, albeit one that clearly ran five minutes fast. But it was different on weekends. The alarm was switched off and we could all sleep in. And that's all of us, for there wasn't a peep out of Maxi on Saturdays and Sundays. In fact she slept in longer than we did – great evidence that dogs can count. Now, you could argue that the dog had observed our behaviour and registered the fact that we sleep longer on weekends. But you can rule out that line of reasoning because during the week she always woke us before our alarm went off, and while we were all still asleep. In the same situation on the weekend, however, she didn't do that. We never did figure out why she stayed in her basket then and slept even longer than the rest of us.

Just for Fun

ARE ANIMALS CAPABLE OF HAVING FUN? That is to say, are they capable of doing things that have no particular purpose other than to bring them pleasure and happiness? I think that's an important question, because the answer helps us decide whether animals experience positive feelings only when they perform tasks that promote the survival of the species (such as pleasure during sex, which creates offspring). If that were the case, then pleasure and happiness would be by-products of purely instinctive programming that ensures certain behaviours are engaged in and rewarded. In contrast, just by remembering happy experiences, humans can relive the emotions that went with them and enjoy them over and over again. Free-time fun belongs here, such as a holiday by the sea or winter sports in the mountains. Might this be the crowning glory that sets us apart from animals? But then those tobogganing crows come to mind. An Internet video shows one of these birds sliding down the roof of a house. The crow has found a lid from a plastic container. It carries it up to the highest part of the roof, places it on the slope and then jumps onto it to slide down. No sooner does the bird reach the bottom than it goes back up for its next ride.[35] The point? Apparently none. The fun factor? Probably the same as when we jump onto the wooden or plastic object of our choice and careen down a hill in the snow.

Why would crows expend energy on such a pointless activity? After all, tough evolutionary competition calls for the elimination of all non-beneficial activities and ejects from the race any animal that is not sufficiently rigorous in this respect. And yet it's been a long time since we've paid any attention to this seemingly absolute rule. At least in wealthier countries, we have energy to spare and we can afford to use it to enjoy ourselves. Why should it be any different for an intelligent bird that has set aside sufficient food for the winter and can devote some of those calories to fun and games? Clearly crows, too, can convert surplus resources into mindless fun and conjure up happy feelings whenever they want.

So what about dogs and cats? Anyone who lives with these animals can tell stories about how they love to play. Our dog Maxi liked to play tag with me around our forest lodge. Because she knew she could run much faster than I could, she always gave me a chance to catch her so that the game didn't get boring. She'd run big circles around me, every once in a while dashing towards me. Then, just before I caught her, she'd sidestep and I'd miss. You could tell just by looking at her how this game delighted her. I really enjoy looking back to that time, and yet I'd rather find other examples as evidence of completely pointless play – meaning pointless in a positive way – because Maxi probably used this game to strengthen our relationship. And it's true that any playful activity within a group can act as social glue and therefore serve an evolutionary purpose. Energy invested in cohesion promotes groups that are particularly resistant to external threats.

So let's take another look at crows. There are lots of reports of crows that tease dogs. They stalk them from behind and nip them on the tail. Of course the dog spins around too slowly to catch the bird, which soon starts the game all over again. This is not a case of creating social cohesion, and it's not a case of the bird honing some survival skill, either. After all, escaping from spinning dogs is not a necessary part of its behavioural repertoire. No, what's going on here seems to be something completely different. The crow can clearly put itself in the dog's place and realises that the dog will always be too slow and will therefore get annoyed. And that's what makes it so much fun to tease it over and over again, gleefully anticipating its reaction. Lots of crows enjoy doing this, as any number of Internet videos attest.

Desire

SEX IS NOT AN AUTOMATIC PROCESS for animals, although you'd be forgiven if, after reading scientific papers on 'mating', you concluded that it is an act completely devoid of emotion. Hormones feature, and they trigger instinctive behaviours that animals are incapable of resisting. But is it any different for people? I just have to think back to the couple I came across years ago in the woods. All I wanted to do was check out whose car was parked in the undergrowth, when two beet-red faces appeared from behind the bonnet. I knew both the man and the woman. They came from neighbouring villages and at the time were married to different partners (and still are). They quickly adjusted their clothing, climbed into their cars without saying a word and disappeared. Clearly they didn't want to risk their marriages and had therefore chosen a supposedly secluded spot for sex. Even though there remained a risk of dire consequences for their personal lives, both of them had succumbed to desire. I think this is a good example of how we, too, are at the mercy of our instincts.

The trigger for such behaviour is a cocktail of hormones that gives rise to feelings of exquisite satisfaction and joy. But why is that even necessary? If living beings are supposed to mate, they could mate as instinctively as they breathe. After all, our bodies don't release supplements of drug-like substances to trigger every breath. Mating is special because during copulation all species abandon themselves to a state of helplessness. Sadomasochistic animals such as snails ram chalky darts into each other's bodies to stimulate their partners. Male peacocks and grouse attract their hen's attention with a fan of erect tail feathers before jumping on her. Insect couples hitch aerial piggyback rides. Male toads in the raptures of love clamp down on their female underwater; sometimes multiple males pile up on top of one another and don't let go, holding the female down for so long that she drowns.

Goats, which behave in many ways like deer, undergo a somewhat more elaborate ritual every year in late summer. At that time of year our billy goat, Vito, turns into a monster of stink. To please the ladies, he perfumes his face and forelegs with a very special fragrance: his own urine. And he doesn't just spray the yellow liquid onto his skin; he also sprays it into his mouth. What would make us gag obviously doesn't fail to have the desired effect on the female goats. They rub their heads over his coat to absorb the smell. This clearly stimulates hormone production in all involved, and the fires of desire are lit. The billy goat keeps testing with his nose to see if one of the does is ready to let him in. He does this by driving her around the pasture while bleating with his tongue hanging out, which actually looks a bit ridiculous. If the lady of his dreams stands her ground and squats to pee, he shoves his nose into the stream. Then he snorts and curls back his upper lip to check her hormone levels, to see if this will be his lucky day. After many days of testing, the does finally allow Vito a few seconds of bliss.

Let's get back to the question of why there has to be any sort of hormone-fuelled emotional reward. The back-story is that mating is dangerous. The prelude to mating, when males often draw attention to themselves, doesn't just attract females. Indeed, hungry predators are also grateful for an eye- or ear-catching sign pointing the way to a tasty meal. And quite a few males of many different species do, in fact, step off their woodland stage directly into the stomach of a bird or a fox. The actual act of mating is even more dangerous. While it's going on, both participants are pressed tightly together for a few seconds, and sometimes even for many minutes, and are hardly in a position to flee from an attack.

We don't know if animals make the connection between mating and offspring, yet what other reason could there be for them to take this risk? Only one comes to mind: the strong, addictive feeling of orgasm that causes them to throw caution to the wind and abandon themselves to pleasure. Therefore I believe, without a doubt, that animals experience intense emotions when they have sex. And there is another strong indicator that this is so. Quite a few animals have been observed pleasuring themselves. Deer, horses, wild cats and brown bears – all have been seen either laying a hand (that is to say, a paw) on themselves or taking advantage of natural aids such as tree trunks. Unfortunately, there are not many reports, let alone any research, on the subject, perhaps because masturbation is considered taboo.

Till Death Do Us Part

DOES IT MAKE SENSE TO THINK of animal pairings as marriage? According to the dictionary definition, marriage is a legally recognised lifelong partnership between two individuals. Wikipedia states that marriage is 'a socially or ritually recognized union or legal contract between spouses'.[36] There is no such thing as legal recognition for animals, but there is such a thing as a close lifelong partnership. The raven is a particularly moving example. It is the largest songbird in the world and, by the middle of the twentieth century, it was almost extinct in Central Europe. It was alleged that ravens killed grazing animals as large as cows. Today we know that this is nonsense: ravens are the vultures of the north, and they seek out dead or at most dying animals. Yet they were ruthlessly hunted and killed, using both weapons and poison.

Historically, campaigns to get rid of undesirable animals have met with mixed success. In the twentieth century people wanted to eradicate the red fox because it can carry rabies. Foxes were shot on sight (and they still are), their dens were dug up and any young found in them were destroyed. Poison gas pumped into their underground accommodation was a convenient way to kill them. But despite these measures, the red fox has survived, because it is very adaptable and reproduces prolifically. And, most importantly, red foxes change their sexual partner. In contrast, ravens are faithful souls and stick with their mate for life. In this respect, we can indeed talk of a true animal marriage. During the extermination campaign, the birds' fidelity was their undoing. If one bird in a pair was shot or poisoned, the other often didn't seek out a new partner, but from then on flew its circuits in the sky alone. The large number of single birds contributed no young, which hastened destruction of the species. Today, ravens are strictly protected in Central Europe and once again populate their ancestral range. I still remember a trip Miriam and I made to Sweden with our children. As we paddled our canoes over a deserted lake we often heard ravens call, and I was entranced. I was so excited when a few years ago I heard these birds for the first time in the forest that I manage in Hümmel. Since then, I think of these animals as symbols for how Nature can heal from our crimes against it, and how destruction of the environment need not be a road of no return.

Monogamous animals are common, especially among birds. There are several species that behave like ravens, although none quite as rigorously. They remain true to the same partner, at least within a single breeding season. There's the white stork, for instance, although what the stork remains true to is actually its nest from one season to the next, and often the only reason partners meet up again is because both head back to their old nest every spring. Sometimes, however, things can go awry, as reported by an employee at the Heidelberg Zoo. One spring, a stork built a nest up high with a new female. His former partner had obviously gone astray during migration – but then, in the middle of this cosy togetherness, she belatedly appeared. Now the male stork found himself in a quandary. In order to be fair to both females, he built a second nest and then ran himself ragged providing for both families.[37]

But why aren't all species of birds so loyal? And what does being loyal really mean? Just because chickadees and other species don't mate for life doesn't mean they are disloyal partners. The reason for a bond that lasts for just one season lies in the average lifespan of the birds. Whereas ravens – even in the dangers of the wild – live for more than twenty years, for other species, mostly the smaller birds, it's all over after less than five. If you mate for life and the likelihood your partner will go missing is very high, then there would soon be mostly singles flying around. And because that would be exceedingly bad news for the survival of the species, for birds with short life expectancies there's a new roll of the dice every spring to see who partners up with whom. That's when the birds can see who survived the winter and migration. Great tits and robin redbreasts waste no time grieving over a mate from the previous year that never returns.

And what about mammals? There are only a few that hook up like ravens do. Beavers are one such example. They find a life partner and remain with the same animal for up to twenty years. Their children don't move out immediately, and the young continue living with their parents in comfortable earth lodges close to the water until they are two years old. Most other species are somewhat challenged in the relationship department, at least when it comes to the opposite sex. With red deer, it's a case of 'to the strongest go the spoils'. Once a mighty stag has routed his rivals, he satisfies himself with the harem of does until he, in turn, is driven off by an even stronger male. It's clearly all the same to the does. They allow themselves to be mounted by any young buck that tries his luck when the alpha male is not paying attention. Raising fawns is purely women's work anyway, because by the time they are born, their fathers are already off wandering through the woods in male-only groups.

What's in a Name?

IT SEEMS OBVIOUS TO US that we can call each other by name when we communicate. In large societies, each person we want to address has a personal name that we can use when we want to get their attention. Whether we're using email, an online app or the phone or are having a face-to-face conversation, without this direct form of address we'd be lost. We're conscious of how important this is when we meet someone we've been introduced to before and we forget their name. Is naming a particularly human habit or is there something similar in the animal realm? After all, all social creatures face the same issue.

In mammals, there is a simple form of calling between mother and child. The mother utters a call using her normal voice. The child recognises the call and responds with a clear vocalisation of its own. But is this really an exchange of 'names' or is it merely voice recognition? The argument in favour of the latter is that these special mother–child 'names' seem to fade with time. Once the young animals are grown up and weaned, their mother no longer responds to them. What meaning does a name that you've given yourself have, when no one reacts to it? Does a temporarily meaningful call deserve to be called a name?

Even if we disallow such calls, science has discovered some instances of genuine personal naming in the animal kingdom. It's no coincidence that this naming happens, once again, with ravens. Their close societal bonds provide an ideal background for our investigation, because ravens cultivate lifelong relationships, not only between parents and children, but also among friends. Naming calls are the way to go, if you want to communicate over long distances and identify which individual is doing the talking. These inky black birds can master more than eighty different calls – a raven vocabulary, if you will. Amongst them is a personal identification call that a raven uses to announce its presence to other ravens. But is this call really a name? That would only be true, in the way we use the word, if other ravens also 'addressed' the speaker using its personal identification call – and that's exactly what ravens do.[38] They remember the names of other ravens for years, even if they've had no contact. If an acquaintance appears in the sky and calls his or her name from afar, there are two possible responses. If the returning raven is an old friend, the other ravens answer in high-pitched, friendly voices. However, if the raven is unpopular, the greeting is low-pitched and brusque. Similar observations have been made concerning human greetings.[39]

It's quite difficult to discover the names animals use for each other. It's much easier if we call them using a name we've chosen for them and see if they react. People who own a single pet now face another difficulty. How do we know that, for instance, our dog Maxi doesn't hear her name and interpret it to mean simply 'Hi' or 'Come here'? It would be easier to tell if you owned more than one dog. But let's return to the intelligent pig. Researchers have studied pigs for precisely this attribute. The reason for the research was the widespread pushing and shoving in modern pens. Farmers used to pour feed into a long trough so that all the pigs could eat at the same time. Today, feed is distributed automatically with the help of computers that calculate the amount for each individual pig and, because the equipment is expensive, there's not enough money for many machines, so not all the pigs in one pen can eat at the same time. They have to wait their turn and, if their stomachs are rumbling, pigs are as impatient as we are. They jostle each other in the queue, sometimes even injuring one another.

To help bring some civility back into the process, researchers from the 'pig team' of the Friedrich-Loeffler-Institut (the German Federal Research Institute for Animal Health) tried to teach manners to pigs at an experimental farm in Mecklenhorst in Lower Saxony. Here, using small school 'classes' of eight to ten animals, they trained yearlings to respond to individual names. The young porkers were particularly good at catching on to three-syllable female names. After one week of training, the pigs went back into a pen in a larger group. And now meal times got really interesting. Each pig was called up by name when it was its turn. And it worked. As soon as 'Brunhilde' rang out from the loudspeaker, the only pig that sprang up and raced to the trough was the one that had been called, while all the others continued doing whatever it was they were doing, which for most of them was simply snoozing. The measured heart rate of the other pigs did not increase, and the only pig that registered a higher pulse was the one that had been summoned. The new system had a success rate of 90 per cent. It is one way to bring peace and order to the pens.[40]

But does this heart-warming discovery have a wider significance? The ability to associate oneself with a particular name presupposes self-consciousness. And that is one step above consciousness. For whereas the latter merely implies thought processes, with self-consciousness we're talking about recognising who you are – having a sense of self. To test whether animals possess this ability, science came up with the mirror test: animals that recognise that the reflection in the mirror is not another animal, but their own image, are supposedly self-aware. The inventor of this experiment was one Gordon Gallup, who painted a coloured spot on the forehead of anaesthetised chimpanzees. Then he placed a mirror in front of the drugged animals and waited to see what happened when they woke up. No sooner did the apes glimpse their reflections through eyes still blurry with sleep than they began to try and rub the colour off. Clearly they understood immediately that they were the ones looking out from the gleaming glass. Ever since, this test has been considered proof that those animals that pass it possess self-consciousness. (Incidentally, children don't pass this test until they are about eighteen months old.) Primates, dolphins and elephants have all passed the test since it was first introduced and, accordingly, have risen in the eyes of researchers.

They were surprised when crows also recognised their own reflection, as did magpies and ravens. In response to corvids' intelligence, scientists have started referring to these birds as 'feathered apes'.[41] There were no discoveries of this kind for a long time but then, suddenly, pigs began popping up in scientific papers. Pigs? Yes, they too passed the test. Unfortunately they didn't score a pet name – 'factory-farm apes' comes to mind – for if they had, how could anyone have continued to treat pigs as inhumanely as they are treated today? People don't even credit these intelligent animals with feeling pain, as is evident from the fact that in Germany it is legal to castrate piglets only a few days old without anaesthetic. And it will continue to be legal until 2019, because it's quicker and cheaper that way.

But let's return to the mirror. Pigs know how to use mirrors for more than just self-contemplation. Donald M. Broom and his team at the University of Cambridge placed food behind a barrier. Then pigs were positioned so that they could only see the food using a mirror placed in front of them. Seven out of eight pigs took just a few seconds to realise that they had to turn round and go behind the barrier to reach the tasty treats. To do this, they not only had to recognise themselves in the mirror, but also had to consider spatial relationships in their surroundings, and their own place in them.[42]

Despite these results, we should be careful not to place too much store by the results of the mirror test, particularly when it comes to animals that do not pass it. First of all, when dogs look at their reflection and do not react, that may not mean anything at all. How are we to know whether the spot on their face bothers them in the slightest? And even if it does, perhaps they don't know how the mirror works. Perhaps they think it's a colourful picture or, at most, a video like the ones we watch on a television.

Let's return to naming, and now those Canadian squirrels take centre stage again. When investigating cases of adoption, researchers discovered that these arboreal imps adopted only related young. But how did they know which ones were their nieces, nephews or grandchildren? Researchers at McGill University suspect that adults' vocalisations have an important role to play here. Every squirrel has a distinctive call, and the solitary creatures use these calls to recognise one another. After all, they don't see each other very often because their territories rarely intersect, so the only way they can communicate is by sound. What is even more astounding is that some animals come looking, when they stop hearing their relatives' calls. This means they have to leave their own territory and enter an unfamiliar one. Does that make them nervous? We can only speculate about that, but we do know that when they come across orphaned young during their forays into alien territory, they take the helpless young into their care.[43]

Science is only just beginning to explore this subject, like so many others. As I've explained, naming is an advanced form of communication that many animals have mastered. Even fish, which we think of as silent creatures, are in on this skill, but the only thing we know right now is that they use the sounds they make to find a partner or to defend their territory.[44]

Grief

RED DEER ARE SOCIAL ANIMALS. They congregate in large herds and enjoy being in a crowd. That being said, there is a definite difference between the sexes. After they turn two, male red deer become restless and wander far afield. They get together with others of their sex, but they stay only loosely connected. As they mature, they become solitary and prefer to be on their own, occasionally tolerating the presence of a younger male in their vicinity. German hunters call these hangers-on 'adjutants'.

A female red deer is much more constant. Her herd is a tightly knit community led by a particularly experienced alpha doe. The lead doe passes along to the younger females established behaviours passed down to her by her predecessors, including the use of decades-old travel corridors between territories. These are paths the herds use to reach lush summer meadows or protected winter quarters. When danger threatens, the frightened females look to their leader. She's the first to know what to do, because she can remember similar situations and potential predators. These dangers don't just come from animals. For example, I have often observed a herd of deer leaving an area at the start of a shoot. A few blasts on the trusty hunting horn signal that it's time for the deer to get going. The call to action warms the hunters' hearts before they set off – and warns the lead deer that it is time to depart. This proves deer can remember the specific sequence of notes, even though more than a year has passed since the last hunting season.

Besides age and experience, an alpha doe needs one other attribute to prove she deserves her position: offspring. Offspring are essential to show that the doe can take responsibility not only for herself, but also for others. Some wildlife researchers attribute the fact that the other deer in the herd follow her to happenstance. The way they see it, because the deer feel comfortable only when they are in a group, and as the older doe is leading her fawn anyway, they join her more or less randomly merely because two animals happen to be moving off in the same direction. I, however, am convinced that the members of the herd are well aware that a particularly experienced doe is taking the lead. She decides first, and the others are happy to defer to her decision. But researchers object to this, saying that older animals are more vigilant and that's why she's the first to react when it's time to flee. It's hardly surprising that the others choose to follow her, to be on the safe side. What's going on is simply passive leadership and not active guidance.[45] I don't believe that, either. It's true that does don't fight for dominance in the herd and resolve issues of hierarchy quietly in ways we don't yet recognise. However, if leadership were merely a matter of chance, the does would follow one female one day and another the next – perhaps even an especially fearful doe, one that is not only young and inexperienced but also distinctly nervous, making her particularly prone to flight. But true leadership is characterised by something completely different: the ability not to get worked up unnecessarily, because an animal that panics too often has less time to eat and therefore less energy to ensure survival. It is because the other does recognise the experience that comes with age that they all quietly agree amongst themselves that the lead doe is the one they will follow.

There are times, however, when disaster strikes the alpha doe – for example, when her fawn dies. In earlier times, the cause of death was usually disease or a hungry wolf. These days, it is often a blast from a hunter's gun. For deer, the same process is set in motion as for us. First, unbelievable confusion reigns, and then grief sets in. Grief? Are deer even capable of experiencing an emotion like that? Not only can they, but they must. Grief helps them to say goodbye. The bond between doe and fawn is so intense that it cannot be severed from one moment to the next. The doe must slowly accept that her child is dead and that she must distance herself from the tiny corpse. Over and over again she returns to the spot where her child died and calls for her fawn, even if the hunter has carried it away. A grieving alpha doe endangers her kin by remaining close to the scene of death, which also means close to danger. What the herd needs to do is move to a safer place, but departure is delayed until the bond between mother and fawn finally dissolves. There is no question that, under such circumstances, it is time for a change in leadership, and this happens without any struggle for dominance. Another similarly experienced doe simply steps forward and takes over leadership of the group.

If it's the other way round and the alpha doe dies, leaving her fawn behind, the remaining does show no mercy. There is no question of an adoption. Quite the opposite. The orphaned youngster is frequently banished from the herd, perhaps because the other does don't like dynasties. Left to fend for itself, the fawn has little chance of surviving and usually doesn't make it through the next winter.

Shame and Regret

I NEVER WANTED TO HAVE HORSES. I thought they were too big and too dangerous, and I was never interested in riding. That is until the day we bought two of them. My wife, Miriam, had long dreamed of living with horses, and there was enough pasture for rent close to our forest lodge. So when a horse owner who lived not too far away wanted to sell his animals, the ideal moment seemed to have arrived. Zipy, a Quarter Horse mare, was six and broken for riding. Her companion, a four-year-old Appaloosa mare called Bridgi, was considered unrideable because she had a diagnosed problem with her back. That was perfect, as far as I was concerned. It had to be the two of them because herd animals should not be kept alone, and the fact that only one of them could be ridden was just fine with me, because I was off the hook when it came to riding.

But then everything changed. Our vet examined both animals and decided there was nothing wrong with Bridgi after all. What was to stop her from being ridden as well? Nothing. And so Bridgi and I began to learn together, with instruction from a riding coach. As I rode her, but even more as I cared for her daily, I developed a very close relationship with the mare, until my fear of horses disappeared completely. Moreover, I learned how sensitive horses are and how they react immediately to the slightest cues. If Miriam or I wasn't paying close attention or appeared to be tense, they ignored our commands or shoved inconsiderately when we were feeding them. It was the same when we were riding them. The horses knew just by how tense our bodies were whether our cues – for instance, a small adjustment of weight in the direction we wanted them to go – were to be taken seriously or not. Conversely, in time we also learned how to pick up on very small cues from Zipy and Bridgi. And while we were handling the horses, we discovered the depths of their emotions.

For example, our horses have a highly developed sense of fairness, which shows itself in all kinds of situations. It's particularly marked and most easily observed at feeding time. Zipy, at the age of twenty-three, no longer digests pasture grass very well and would slowly waste away if we didn't step in to improve her diet. So every day at noon she gets a large scoop of a concentrated grain-based feed. If Bridgi, three years her junior, has to watch, she becomes petulant. She dances around and lays her ears back – a guarded threat. In short, she sulks. So we throw her a handful of feed, which we spread out in a long line on the grass. Because she has to separate the feed from the blades of grass, she spends just as long eating as her older companion in the manger with her larger portion. And that sets the world to rights, as far as Bridgi's concerned.

We noticed the same thing with training. The horses clearly enjoy being ridden in the small riding ring. And not just because of the exercise. They have plenty of that, because they spend the whole year out in a large pasture. No, what the horses really enjoy is the attention we pay them when we practise different moves, and the praise and pats we give them when something works out well.

As we spent time with our horses, we noticed the stirring of another emotion: they can feel ashamed or embarrassed just as we can. Despite her twenty years, the lower-ranking Bridgi sometimes behaves like a silly and naughty young thing. When she's in one of those moods, she won't come directly to us when we call, but prefers to take another turn around the pasture at a gallop, or tries to grab something before we've given the command: 'Time to eat!' Then we reprimand her, which we do by making her wait just a little bit longer for her food until she starts behaving herself again. Normally she takes the rebuke in her stride, but if the older Zipy is watching, she turns her head away sheepishly and starts to yawn. You can tell just by looking at her how embarrassed she is. Or, to put it another way, Bridgi is ashamed of her girlish behaviour.

Humans too mostly display shame in front of other people; in fact it is somebody else's presence that makes things awkward for us. Clearly it's the same for horses, and I believe that many social animals experience similar feelings. The underlying reasons for these emotions have not been researched in animals, but they have been examined in people, and this gives us an idea why shame and embarrassment exist at all: the person in question has transgressed social rules. They turn red and lower their gaze, to signal submission. The other members of the group see their misery and, as a rule, feel sympathy for them, which means that the person who made the misstep is usually forgiven. In essence, shame and embarrassment function as a kind of act of contrition – they are mechanisms for asking for forgiveness. It is still mostly assumed that animals cannot feel ashamed or embarrassed, because to do so an individual must be in a position to think about their own behaviour and the effect it has on others.[46] I am unfortunately unaware of any current research on this topic, but there is a related emotion that I can report on, and that is regret.

How often in life do we regret making a bad decision? Regret is an emotion that generally protects us from making the same mistake twice. It's a perfectly sensible response, because it stops us wasting energy by repeating dangerous or pointless behaviour over and over again. And if it makes that much sense, then it also seems obvious to look for such an emotion in the animal world. Researchers at the University of Minnesota in Minneapolis observed rats, with this in mind. They built a special 'restaurant row' for rats – a ring with four spokes leading to four different feeding zones. When a rat came to the entrance of one of the spokes, a sound indicated how long the wait for food would be: the higher the sound, the longer the wait. And now the rodents began to act like people. Some lost patience and went on to the next spoke, in the hope that there they would be served more quickly. Sometimes, however, the sound was even higher there, meaning the wait time would be longer. Now the animals looked wistfully back in the direction of the spoke where they had just been, but they also grew more determined not to change zones again, but to wait longer for their food. People react in similar ways – for example, when we switch queues at the supermarket checkout and realise we've made the wrong choice. The researchers detected patterns of activity in the brains of the rats similar to those in our brains when we mentally replay our shopping predicament. That's what makes regret different from disappointment. The latter kicks in when we don't get what we were hoping for. In contrast, we feel regret when we also realise there could have been a better outcome. And researchers Adam P. Steiner and David Redish discovered that rats can clearly come to the same conclusion.[47]

If rats exhibit these feelings, doesn't it make sense to look more closely for such feelings in dogs? After all, almost every dog owner knows that dogs are sorry when they behave badly, which they demonstrate by pulling a typical pitiful puppy-dog face when they're told off. Our Münsterländer, Maxi, knew very well when she had done something wrong and I scolded her. She would put her head down and look up at me, as though it was all terribly distressing and she was asking me for forgiveness. And exactly this behaviour has been put to the test by researchers. Bonnie Beaver from Texas A&M College of Veterinary Medicine came to the following conclusion: dogs give their owners the typical puppy-dog look because they learn what their owners expect when they scold, which means the dogs are reacting to the scolding and not to their guilty conscience.[48] Alexandra Horowitz from Barnard College in New York came to the same conclusion. To do this, she asked fourteen dog owners each to leave their dog in a room with a bowl full of treats, after giving the dog a strict warning not to touch anything. All the owners were then told that their dogs had eaten the treats. The result: although some of the dogs had in fact obeyed the order, all of them made the puppy-dog face as soon as they were scolded.[49] And yet that doesn't necessarily mean that dogs only pretend to feel sorry for their actions. If the scolding comes right after the act, dogs connect their owner's reaction with their behaviour, and then their look probably really does express the remorse we ascribe to them.

Let's return once more to the feeling of fairness, for that definitely exists in the animal kingdom, and not just among horses. If you live in a social group, things need to be fair. According to the dictionary definition of the term 'justice', it means that every member of a community should be treated equally. If they aren't, resentment quickly bubbles to the surface and, if this resentment is constantly fed by further injustices, it can lead to violence. In human communities, laws are supposed to protect everyone's interests. However, emotions such as shame, when we behave badly, and happiness, when we behave well, are considerably more important than the law when it comes to our daily dealings with each other. How else could fairness function within the privacy of our own four walls, within our own family?

I've already written about how our horses feel ashamed and have an innate sense of fairness. Of course this is a personal observation I've made and not the proven result of a scientifically controlled experiment, but such experiments do exist for dogs. To prove that dogs have a sense of fairness, Friederike Range's team at the University of Veterinary Medicine in Vienna had two dogs that knew each other sit side by side. All the dogs had to do was follow a simple command: 'Give me your paw.' When they obeyed, there was a reward. This could vary greatly: sometimes it was a piece of sausage, sometimes only a piece of bread and sometimes it was – nothing at all. As long as the rules of the game were the same for both dogs, everything was fine, and the dogs were happy to play along. To make the dogs feel envious, they were rewarded most unfairly as the experiment progressed. If both offered up a paw, only one got a reward. The more stringent tests rewarded one dog with sausage while the other got nothing, even though it had offered its paw like a good dog. The unfair gift of food to the other dog was eyed suspiciously. It didn't matter whether the other animal got the tastier treat with or without doing anything – at some point the dog that was being unfairly treated got fed up and refused to cooperate. But if the dogs were separated and unable to compare themselves with their companion, then they didn't object and continued to cooperate even when there was no reward. Before this experiment, researchers had observed such feelings of envy and unfairness only in primates.[50]

Ravens also have a strong sense of right and wrong. Researchers discovered this while running experiments designed to explore cooperation and tool use. In these experiments, a small board with two pieces of cheese was put out behind bars. The cheese was attached to a cord, and the ends of the cord were threaded through the bars to two ravens. The birds could only pull the treats within reach if they both pulled carefully on the ends of the cord at the same time. It didn't take these clever birds long to catch on, and the test worked particularly well with couples that liked each other. But what happened with some teams was that after successfully pulling the cheese closer, one of the ravens snapped up both pieces. The bird that ended up with nothing now refused to work with its greedy colleague. Selfish individuals, it seems, aren't popular in the bird world, either.[51]

Empathy

THE COMMONEST MAMMALS IN THE FOREST are also among the smallest representatives of this class of vertebrates: wood mice. They are pretty little things but, because they are so small, they are difficult to observe and therefore many woodland walkers are not that interested in them. I didn't realise how many of these tiny creatures are scurrying around in the underbrush until I had to stand around for a long time in one spot waiting for an appointment with someone interested in our woodland burial ground. Wood mice are omnivores, and they spend their summers under old beech trees, living the good life. There are buds, insects and other small animals aplenty, so they can take it easy and concentrate on bringing up their offspring. But then winter approaches. To avoid the worst of the cold, the mice set up their living quarters at the foot of mighty trees, usually where multiple roots cover the forest floor. The roots create natural hollows, and all the mice need to do is enlarge them a bit for their purposes. Wood mice are social creatures, so a number of them usually share the same space.

When there's snow on the ground, I can sometimes spot the traces of a drama that played out among the roots. A trail of small paw prints leads to the trunk of a beech – a marten on the move. And martens love mice for breakfast. The tracks lead to a root hollow, and now I can clearly see signs of a whole lot of scraping and scratching. Not only did the marten nonchalantly dig up the hidden hoard of mouse food, but it might even have dug up one of the mice. What must that have been like for the other mice? Were they simply afraid of the marten or did they also realise that its activities caused one of their own to suffer?

As researchers at McGill University in Montreal have discovered, they would have been clearly aware of their fellow mouse's suffering. The researchers observed evidence of empathy in the tiny mammals, the first non-primates in which such emotions have been documented. The experiments themselves, however, were anything but empathetic. The researchers caused painful injuries by injecting acid into the mice's tiny paws, or, in another experiment, pressed these sensitive body parts onto hot surfaces. If the mice had observed other mice undergo similar torture, they experienced considerably more pain than if they went into the experiment unprepared. On the flip side, the presence of another, less traumatised mouse made it easier for the test subject to endure the pain. What was important was how long the mice had known each other. There were clear signs of empathy if the animals had been together for more than fourteen days, which is typical for wild wood mice in Central European woodlands.

But how do mice communicate amongst themselves? How do they know another mouse is suffering and experiencing a private hell? To find this out, the researchers blocked the mice's senses one after the other: first sight, then hearing, smell and, finally, taste. And although mice like to communicate through smell and make shrill ultrasonic calls when alarmed, surprisingly enough, in the case of empathy, it is the sight of suffering companions that triggers their response.[52] So when a marten fishes a wood mouse out from a cosy root-hollow in winter, the other mice must endure their own few seconds of terror. We don't yet know how long such feelings of empathy last. So when I come across the marten's tracks in the snow, I have no idea whether their sympathetic responses and the upset they cause are still distressing the tiny inhabitants below.

But how does empathy work with mice that have only recently joined a group and are therefore not yet fully integrated? Clearly, it is felt considerably less intensely, and in this respect mice are no different from humans, which the researchers at McGill University also discovered. They compared the empathetic behaviour of students and mice, and concluded that empathy for family members and friends is much more pronounced than empathy for strangers. The reason was the same for all the experimental subjects – stress. Stressed individuals are less affected by the suffering of others. Strangers are often the cause of this stress, and the sight of them releases the hormone cortisol. To verify this, researchers carried out another experiment, this time using a drug that blocked the production of cortisol in both students and mice, and feelings of empathy increased again.[53]

And domestic pigs yet again show their feelings when it comes to empathy. In this case, I'm thinking of the experiments of Dutch scientists at Wageningen University and Research Centre in charge of the experimental pens at the Swine Innovation Centre in Sterksel. Here they played classical music to the pigs. Don't worry, the researchers weren't trying to find out if the animals are fond of Bach. Rather, they got the pigs to connect the music to small rewards, such as chocolate-covered raisins hidden in the straw. Over time, music triggered particular emotions in the pigs in the experimental group. And now things got interesting, for other pigs were added that had never heard such sounds and therefore had no idea what they meant. Despite this, they experienced the same emotions that affected the musical pigs. If the musical pigs were happy, the newcomers also played and jumped around; in contrast, if the musical pigs were so scared that they urinated on themselves, the newcomers caught the feeling and exhibited the same behaviour. Pigs clearly can experience empathy. They can pick up on the emotions that other pigs are feeling and experience those feelings themselves – a classic expression of empathy.[54]

And what do things look like, emotionally speaking, between different species? It is clear that we react to the suffering of other species. Otherwise, why would we be so disgusted by photographs of naked, bleeding chickens in dark battery cages or of apes with exposed brains attached to research equipment? A particularly moving example that animals, too, are capable of empathy across species-lines comes from the Budapest Zoo. Zoo visitor Aleksander Medveš was filming the brown bear in its enclosure when suddenly a crow fell into the surrounding moat. The bird began to weaken as it thrashed about and was in danger of drowning, when the bear intervened. It carefully took one of the bird's feathers in its mouth and pulled the bird back to land. The crow lay there as though petrified, before it pulled itself together. The bear took no more notice of this fresh morsel of meat, even though it was potential prey. Instead, it turned its attention once more to its meal of vegetables.[55] A freak occurrence? Why would the bear do such a thing, when it clearly had nothing to do with either the urge to eat or the desire to play?

Perhaps it would help answer the question of whether a species is capable of empathy if, in addition to direct observation, we were to take a look inside the brain. To locate empathy, researchers check for the presence of mirror neurons. These specialised cells were discovered in 1992, and they have an unusual quality. When you engage in a particular behaviour, normal nerve cells in your body fire electrical impulses. Mirror neurons, in contrast, become active when someone else engages in the same behaviour. That is to say, mirror neurons react as though it's your body that is affected. A classic example is yawning: when your partner opens his or her mouth to yawn, you feel the need to yawn as well. It's more enjoyable, of course, when you allow yourself to be infected by someone else's smile. More serious situations make this even clearer. If a member of your family cuts their finger, you suffer along with that person as though you had injured yourself, because similar nerve cells in your brain respond. However, these mirror neurons function only if they've been trained early in life. People who have loving parents or other caregivers practise mirroring emotions and strengthen these neurons from an early age, while the capacity to feel empathy wastes away in people who are denied early exposure to this skill.[56]

Mirror neurons, then, are the hardware of empathy, and so what would make more sense than to look and see which animals possess these cells? According to the latest research, all we know so far is that apes possess mirror neurons. We still have to find out which other species are like us in this respect. Scientists often publicly speculate that we can probably expect surprises here, too. They assume that all animals that live in herds or large groups possess similar brain mechanisms, because social units function only if individuals can see things from the perspective of others in the group and feel what they are feeling. I can just see the goldfish in the chapter 'Anybody Home?' waving its fin at us again. As an animal that travels around in a tightly knit group, it's on board with this idea – or at least swimming alongside the boat.

Altruism

CAN ANIMALS ACT SELFLESSLY? Selflessness is the opposite of selfishness, a characteristic that, in evolutionary terms (only the fittest/best survive) is not inherently negative. However, if you live in a community, a certain level of selflessness is needed for it to function, at least when this characteristic is defined in such a way that it doesn't necessarily require free will. Under this definition, many animals act selflessly – even bacteria. For instance, individual bacteria that are resistant to antibiotics release indole, a substance that serves as an alarm signal. Immediately, all the other bacteria in the area take protective measures. And this means that even those that have not mutated to become resistant to antibiotics are able to survive.[57] This is a clear case of selflessness, but, according to current scientific opinion at least, it's doubtful that free will comes into play.

As far as I'm concerned, altruism is only meaningful when you have to make a real choice, when you have to consciously and actively sacrifice something in order to help another. In the final analysis, we can't be sure when this is the case with animals, but we can get closer to the issue by starting with the more intelligent ones. Birds belong in this category, and we can observe them behaving altruistically all the time. When an enemy approaches, for example, then the first great tit to notice the danger gives a warning call. Now all the other tits can fly away to safety. The bird that makes the call, however, has put itself in danger by drawing attention to itself. Of course it can also try to fly to safety, but it is now highly likely that the attacker will catch it, instead of one of the other tits. Why would the tit run such a risk? In strictly evolutionary terms, it makes no sense, because as far as the species itself is concerned, it makes absolutely no difference which bird is eaten. However, in the long term altruism means not only giving but also taking, and that can mean future advantages for sympathetic and generous individuals, as Gerald G. Carter and Gerald S. Wilkinson of the University of Maryland observed in, of all things, vampire bats.

These South American bats come out at night and bite cattle and other mammals so that they can lick up the blood that flows from the wounds. If they're going to eat their fill, however, they need experience and luck when it comes to finding cattle and making sure their victims don't move. Unlucky or inexperienced bats often go hungry, but only until their well-fed colleagues return to the cave. Here, the successful bats regurgitate part of their meal of blood for their less fortunate cave-mates, so they all get to share in the meal. And I mean all. Surprisingly enough, it is not only close family members that get to eat, but also other bats that are not even distantly related to the one dispensing food.

Why do the bats do this? From an evolutionary standpoint, surely it is the strongest that survive, and giving handouts makes animals weaker rather than stronger. Indeed, finding food takes energy, and animals that feed others as well must use more of it, while exposing themselves to danger more often. In addition, some members of the community could exploit selfless bats and permanently take advantage of their services. But it's not like that, as the two American researchers discovered. You see, the bats recognise one another and know exactly which of their acquaintances are generous and which are not. Those that exhibit especially altruistic traits are the first to be looked after, if they themselves ever run into a string of bad luck.[58]

Does that mean that altruism is selfish? In evolutionary terms, certainly, because the individuals that show these traits have a higher chance of survival in the long term. But there is something else we can learn from the scientists' observations. Clearly the bats have a choice – free will – and they can decide to share or not to share. If that wasn't the case, there surely would be no need for the complicated social network of mutual recognition, attributing particular traits to particular individuals, and the behaviour this gives rise to. Altruism could simply be genetically fixed as another reflex, so there would no longer be any recognisable character differences between the bats. However, selflessness is meaningful only if it happens of the individual's own free will, and vampire bats clearly exercise their ability to make this choice.

Upbringing

JUST LIKE HUMAN CHILDREN, animal children need to be taught how to master the rules of adulthood. Miriam and I were to learn just how important this is when we bought our small herd of goats. The owner of the dairy in the neighbouring village sells only kids, because he needs the milk from the mother goats to make cheese. That means the baby goats have two alternatives: either they end up at the butcher's or they are sold to hobby farms. Our starter herd of four was lucky enough to come to our pasture as a small group. No sooner had they stepped into their fenced-in enclosure than the first little goat panicked, jumped the fence and disappeared into the forest about 700 metres away. We assumed we would never see it again. After all, how would it know where its new home was? Under normal circumstances, its mother would have been by its side. She would have bleated to calm it down until it felt safe. As it was, it didn't have a support system. None at all? What about the other three kids? Although they were all part of the same herd, they clearly couldn't create anything like a safe space for each other. And so the problems continued.

Bärli (the brown runaway) returned, but then the other members of the rascally band took to hanging around outside the fence, and we broke out into a sweat when we had to round them up. Our only hope was that their behaviour would improve after the first young were born. And indeed, as soon as the goats had their first kids, they became calmer and stayed obediently on their designated patch of pasture. Their sons and daughters didn't give us any trouble at all, because they learned from their mother how good goats live their lives out in the field. If a kid stepped too far out of line, it was first called to order with a sharp bleat and, if that was not enough, it was then at the receiving end of a swift butt from its mother's horns. Not a single one of this second generation of goats has ever jumped the fence, and the escape artist Bärli is now our most well-behaved and affectionate goat, regal and relaxed. Of course, getting older helped: Bärli is heavier and therefore a bit less agile than she used to be, but she is also calmer in herself. Having children has certainly given her self-confidence, and she has now risen to become the leader of the herd, which surely brings greater peace to her life.

Doesn't that all sound perfectly normal and natural? I think so, too. However, if you believe that animal behaviour is instinctive and follows genetically fixed programming, the whole situation looks somewhat different. If that were the case, learning would be unnecessary, because every situation would activate an appropriate behavioural response. But this is certainly not what happens, as millions of pet owners can attest. For example, our dogs are not allowed in the kitchen, and they quickly learn that the kitchen is off-limits when we say 'No!' in a very particular tone of voice. For the rest of their lives, they obey the rule, even though it's a rule that makes no sense at all in the animal world.

But let's take another look at teaching wild animals about life out in the woods, and let's start with the smallest: insects. If insects don't grow up in a colony like bees or their relatives, the ants or wasps, then the young tykes are on their own. There is no one to warn them of the dangers of everyday life, and they have to learn about them all by themselves. It's no wonder that most insect young are eaten by birds or other enemies, and perhaps this parentless learning curve is the main reason insects have so many young. Mice also reproduce rapidly, but considerably less rapidly than these tiny creatures. Field mice, for example, have offspring every four weeks, which, in turn, can bear young when they are just two weeks old. However, the little rodents don't walk off and abandon their young to the world. They teach them how to interact with their environment and how to find food. Scientists have researched just how specific this training can be in the case of house mice, which are common where I live. But the research took place far, far away on Gough Island in the raging South Atlantic, thousands of kilometres from the nearest mainland.

Seabirds such as the enormous albatrosses breed here in complete isolation. That is, they did until one day seafarers discovered the island and inadvertently released house mice that had stowed away on their ships. The mice did there what mice do here. They dug holes, ate roots and blades of grass and multiplied magnificently. But then, one day, one of them suddenly got a taste for meat. It must have found out how to kill albatross chicks, which, quite apart from the savagery of the act, is no easy feat, because the chicks are around 200 times larger than their attackers. The mice quickly learned that a large number of them had to keep biting a chick until it bled to death. Especially brutal mice even began to eat the fluffy balls of down while they were still alive.

But back to animal school. The researchers noticed that, for years, chicks were hunted only in particular parts of the island. Clearly mouse parents demonstrated the technique to their children, passing the skill on to the next generation, while other mice in other areas knew nothing about this hunting strategy. This handing down of hunting strategies is also found among many larger mammals, such as wolves. Moreover, young wild boar and deer too are taught by example to follow the paths their family groups have been navigating safely for decades, to move from summer to winter feeding grounds. And that is why such trails are often well trodden and as hard as concrete from long use. I can tell you that animals that learn from older generations avoid an early death. Unfortunately, I cannot tell you whether school in the animal world is any more enjoyable than school in ours.

Getting Rid of the Kids

IT WAS CLEAR TO US, as it is clear to most parents: one day our children were going to have to stand on their own two feet. We brought them up to be independent from an early age, and nature – more specifically, hormones – did the rest. Although puberty entered our house with little fanfare, at this stage in life differences of opinion often led both sides to think that a parting of the ways at some point in the future might be desirable. The educational system played its part as well. After graduating from high school, it's time for college. As there is no college close to our isolated forest lodge, this meant that both our children had to move to Bonn, about 50 kilometres away. Incidentally, the parent–child relationship in our family improved markedly at this juncture, because we were no longer constantly getting on each other's nerves. So what do animals do? With mammals and birds, at least, there is a similarly tight bond between generations that has to be loosened at some point. And most animals have an additional problem: because the majority of them don't have families in the human sense, within a year at most the adolescents have to make way for new babies. So how do animals get their children to move on?

One approach leaves a bad taste in the youngsters' mouth. And I mean that literally. We've experienced this ourselves with our milk goats. If a mother loses her kids in spring as a result of some misfortune, we have to take matters into our own hands and milk her. If we didn't do this, her engorged udder could get inflamed, which would be very painful. To say nothing of the fact that we then get delicious milk that we either pour over our cereal or process into cheese. Did I say delicious? Well, in the first few weeks that's exactly what it is. It tastes smooth and creamy, and it's hard to tell from good cow's milk. However, as spring progresses, the milk tastes increasingly bitter. Eventually, none of us want to drink it any more. This is when we increase the time between milkings, and the goat's milk gradually dries up. It doesn't matter whether it's us or the kids drinking the milk – the taste makes the udder unappetising, and the young goats begin to turn to grass and other things green. That takes pressure off the mother, as the kids no longer depend on her for food. In addition, she allows them only a few seconds at the teat before she gets irritated, lifts her leg and butts them away. Right on schedule, for by the mating season in the autumn she will have had time to top up her physical reserves, both for herself and for the young she will bear the following year.

Although bees don't want to get rid of their children at the end of the summer, they do want to get rid of their men. The drones – gentle, big-eyed creatures without a sting – mooch about the hive all spring and summer. They don't look for flowers. They don't help dry nectar and process it into honey. They don't feed or look after the baby bees. No, they enjoy the good life, getting fed by the worker bees and occasionally flying out of the hive to check if there might be a queen hanging out nearby who is ready for their attentions. If they spot one, they're on her trail right away, but only a lucky few manage to mate with her on the fly. The unsuccessful suitors buzz their way back to the hive, where they allow themselves to be consoled with a sugary meal. They could live like that forever. However, as summer runs its course, so does the worker bees' patience with the moochers. The young queen mated long ago, and her sisters, who have left the hive in swarms, have also been impregnated. Winter is slowly approaching and the precious provisions in the hive have to feed a few thousand overwintering bees – workers that have managed to survive to a ripe old age – and the queen.

Nothing has been put aside for the burdensome drones, and now a hateful chapter in these insects' life cycle begins. During the late-summer drone massacre, the once-pampered little males are rudely grabbed and unceremoniously shown the door. Resistance is useless, even though the drones use their legs to brace themselves against removal. Clearly they don't like this one bit, and their senses are all on high alert. But any drone that puts up too much of a fight is simply stung to death. The workers show no mercy. Any drone left alive eventually dies an agonising death from starvation or quickly finds itself in the stomach of an equally hungry tit.

Once Wild, Forever Wild

A FEW YEARS AGO I got a telephone call from the next village. A worried woman told me she had a fawn in her house and didn't know what she should do. When I asked her questions, I discovered that her children had brought the animal back to the house after they had been playing with it in the woods. Damn! However well meaning that playful gesture might have been, it was a catastrophe for the young animal. In their first few weeks of life, fawns are left lying alone in the undergrowth or tall grass because this is safest for both mother and fawn. A doe with a fawn moves slowly because she has to keep waiting for the youngster. Often the fawn has not yet experienced how serious life can be, and it dawdles behind mum – an ideal target for wolves or lynx. These predators can spot the pair from a long way off and easily grab a meal. That's why mother deer prefer to separate themselves from their little darlings for the first three to four weeks and leave them in a safe place. It is almost impossible to sniff out a fawn. Because they smell of hardly anything at all, their scent doesn't alert predators to their presence. The doe comes by for a quick visit every once in a while to nurse her fawn, and then she takes off again right away. That gives her more time to feed on nutritious buds and new shoots, instead of having to worry about and watch out for the little one all the time.

When some people who have no idea what's going on find a fawn lying there all alone and very still, their almost instinctive reaction is to take care of it. After all, it's hard to imagine what an abandoned human baby would go through, if someone just put it down somewhere and then disappeared. And so, time and again, 'do-gooders' spontaneously step in and take the supposedly orphaned fawn back home with them. However, as they have no idea what to do next, they then call an expert. It's usually about this time that they realise bringing the fawn home was a huge mistake, but by then it's too late to fix the damage. The fawn now smells of people, and it can't be returned to the wood and its mother, because she will no longer recognise her child. Bottle-feeding a fawn is hard work and, at least in the case of male fawns, is also risky, as we shall see in due course.

I find deer a perfect example of how mother love can be expressed in very different ways. Most mammals are like us and seek constant close contact with their offspring; those that behave differently are not heartless, but simply adapted to different life situations. Fawns feel perfectly safe in the first few weeks of life, even without constant contact with their mother. Their behaviour changes only when they are capable of bounding along behind her. Then they stay close to the doe, rarely straying more than 20 metres away from her.

Unfortunately, in these modern times, typical fawn behaviour in the first few weeks of life has other, far more tragic consequences. When danger threatens, a fawn hunkers down, because it instinctively knows that it is extremely difficult to pick up its scent. But often what is threatening it is not a wolf or a hungry wild boar looking for a succulent morsel of meat. What is bearing down on it is a tractor fitted with an enormous mower that is swiftly cutting down grass on hectares of land. And so the crouching fawn ends up under the blades, and in the best-case scenario is killed immediately. Often, however, it stands up just before the blades reach it so that its legs are cut off along with the grass. One way to avoid this would be to have someone walk through the area the evening before with a dog, to signal that this is a dangerous place to be. Then the doe will encourage her fawn to accompany her to a safer place outside the meadow. However, there is often neither the time nor the staff for such rescue operations.

Another example that wild animals are not suited to domestic life or cuddling is the European wildcat. By 1990, it was almost wiped out. Only about 400 animals survived in the central uplands in western Germany, and there was a small remnant population of about 200 individuals in the Scottish highlands. The forest I manage in the Eifel was one of the last remaining refuges for the wildcats, and so I was frequently able to observe one of these diminutive feline carnivores. Since then, their situation has improved considerably. Thanks to conservation and reintroduction efforts, thousands of wildcats now roam once again through the wooded landscapes of Central Europe.

The identifying features of a European wildcat are clear. It's about the size of a powerful-looking house-cat. Its dense coat is tiger-striped with a touch of ochre. Its bushy tail is ringed and has a black tip. The problem is that this is what domestic tabby cats look like as well, even though they are not related to the wild species. Positive identification is possible only by measuring brain size or gut length, or by sending a hair sample off for genetic testing. And, of course, forest visitors don't usually have any of these testing methods at their disposal. Despite that, there are a few clues. Domestic cats have – how shall I say it? – grown a bit soft, and they stalk through the landscape only at warmer times of the year and no farther than a couple of kilometres from their home. As soon as it gets cold and wet in winter, their desire for adventure and the radius of their activity diminish, and they usually don't stray more than 500 metres in any one direction, because house-cats feeling chilly want to be able to return quickly to the warmth of their home. Of necessity, wildcats are much tougher. They don't hibernate or dial down their metabolism, and they have to hunt mice even when there's snow on the ground. Tabby cats out in the snow, kilometres from the nearest village, are therefore certainly wild and free.

Since Roman times house-cats introduced from Southern Europe have vastly outnumbered wildcats. Why, then, have the latter not been extirpated through

cross-breeding? The existence of hybrids proves that
the two species do interbreed. However, that is the
exception. When the two species meet, the tame version
always comes off worse, because wildcats are quick to
live up to their names. And this brings us to the question of whether wildcats make good pets. In rural areas
there must have been many times when individual
animals attached themselves to people, and no doubt
this is still the case. After all, there are enough animal
lovers today who put out food, and the birds that flock
to winter feeders show that wild animals become less
shy around people over time.

Recently in my own village I found out what
happens when a baby wildcat grows up cared for by
humans. A jogger had spied a youngster next to an
isolated path in the woodland that I manage. He resisted
the urge to scoop up the clearly helpless animal and
take it back home with him, and so at first he just
watched it. A few days later he came back to the same
spot, and found the mewing little fur ball still sitting
beside the trail. Clearly, for whatever reason, its mother
had gone missing and, left to its own devices, the baby
cat would have died. This time the jogger picked it up
carefully and brought it back home with him. He checked
with a wildcat recovery centre to see how best to care
for the animal, and the Senckenberg Research Institute
in Frankfurt confirmed, by testing a hair, that this kitten
was 100 per cent pure wildcat.

Because of their shorter intestines, wildcats cannot
tolerate cat food, and so the little wild child was fed

meat. It wasn't long before the family could no longer get close during feeding, because the kitten went on the attack right away. Apart from that, it stayed by the family's side when they went for walks through the meadows and it seemed it might be getting tame after all. But before long the cat became more and more aggressive and began scaring the older house-cat, until it was impossible to keep it any longer. The family eventually dropped it off at a wildlife rehabilitation centre in the Westerwald.

The story shows that many species do not lose their wildness and are therefore not suited for life in the care of humans. There is good reason why every domestic animal today is the result of a long breeding process. And anyone who is just itching to adopt a wild animal, despite the fact that they are ill-suited for domestication, will fall foul of the law as well. Depending on the country, the laws governing conservation and hunting are very strict and state that wild animals can only be kept in exceptional cases with official approval.

And yet some people try to make the impossible possible, especially in the case of the wolf, which is unfortunate, because it was difficult enough to gain sufficient support for its return to Central Europe without any further complications. The wolf poses no danger to humans, because it takes absolutely no interest in us. However, when we imprison it against its will, the situation is very different. So here's the thing. Not only is keeping a wolf against the law, but, just like the wildcat, it remains a wild animal at heart. The next thing people think of is to cross the wolf with a large dog, such as a husky, to get an animal that has the appearance of a wolf but the tractable nature of a domestic dog. But because that's illegal, too, it has led to a black market for these animals, which are imported from the US or Eastern Europe.[59] The high proportion of wolf blood in these wolf-dog hybrids ensures that they are never really tame, and they become stressed when they have to endure living with people. Such forced proximity is ultimately dangerous, because stress leads to aggression.

Kathryn Lord at the University of Massachusetts researched why wolves, which are, after all, highly social animals, are so much more difficult to keep than dogs. Her results show that the answer lies in the way the pups are socialised. Baby wolves are up and about when they are only two weeks old, before they're even able to open their eyes. They can't even hear at this stage. Their hearing develops and becomes functional only after four weeks. So for the next couple of weeks they feel their way around their mother, constantly learning about their world while they are still deaf and blind. They finally gain control of their eyes at the age of six weeks, but by then the little rascals are already familiar with the smells and sounds of their family and their surroundings, and they are firmly integrated into the social life of the pack.

In contrast, dogs are late bloomers, and that's just what they need to be. They mustn't bond too soon with other pack members, because when it comes down to it, their relationship with a human will be the most important relationship of their life. Thousands of years of breeding have delayed the socialisation phase in dogs, and today it starts when they are four weeks old. With both wolves and dogs, the formative period lasts only four weeks. While not all the wolf pups' senses are fully developed at this important time, puppies explore their environment equipped with their full sensory repertoire – and in the final days of this phase of their life, people are part of their environment. This means that whereas dogs basically feel most at home in our company, wolves retain a certain distrust of us all their life.[60] Wolf-dog hybrids clearly retain this fundamental feeling.

In comparison with a fawn, however, a wolf-dog hybrid is harmless. A fawn, dangerous? Not all fawns, but males are indeed potentially lethal for their owners. For in less than a year the adorable dappled baby grows into a mature buck. Deer are loners and do not tolerate any competition in their territory. The affectionate connection established between human owner and deer during the rearing process fades away, and because its caregiver is clearly another deer (at least in the eyes of the buck), it must be a rival. And so it must be forcibly driven off. Any owner who cannot sidestep gracefully, like the buck's natural competitors, will find themselves blindsided by a sharp-horned battering ram. Such behaviour is not the exception, but the norm. Even when the animals are released back into the wild, the danger remains. After all, deer have memories, too, and they don't always avoid people later in life. There was a report in the *Schwarzwälder Bote* in 2013 of a buck that attacked and wounded two women in the evening hours on a playing field in the village of Waldmössingen. It turned out that it had been hand-reared the previous year.[61]

Snipe Mess

As I HAVE ALREADY MENTIONED in the chapter on shame and regret, our horses Zipy and Bridgi are fed a portion of concentrated feed at noon. The energy-rich grain is meant to beef them up a bit, especially our older horse, Zipy. Clearly, horses don't chew their food very thoroughly, because we find a fair number of undigested seeds in their droppings. And now things get rather unappetising, because our 'house crows', which are always hanging about close to the pasture, have their eye on these. They pick the horse apples apart and peck out individual oat seeds. Yummy? I think it looks pretty disgusting, and so the question is: how tasty can such excrement-coated food really be? Do animals have any sense of taste at all? The answer is: they definitely do, but it is adapted to different traditional fare than our palates. (Of course there are also differences when it comes to our taste buds. Just think of the 100-year-old eggs so beloved in China. To many Europeans, at least, these dark translucent eggs conjure up visions of rot and decay rather than culinary delight.)

Our horses offer further proof of animals' sense of taste. We have to worm them two or three times a year. To do this, we squeeze a medicinal paste from a tube into their mouths. Apparently it doesn't taste good at all, because when the two of them notice what's about to happen, they really don't want to be around us. However, the manufacturer has recently made some changes. The de-worming treatment now comes in apple flavour, and horses love apples. Since then, the procedure has gone somewhat more smoothly. Dog owners are also well aware that their pets learn what they like and don't like. For instance, if they change food brands, their four-legged companions sometimes refuse to eat. This was never a problem with Crusty, our French bulldog, who always ate with gusto; however, unfamiliar foods exacted their price, at least for us. A short time after any such new meal, a stink cloud from Crusty's bottom would waft through the air every ten minutes or so, filling the room.

And then there are rabbits, which are slightly more perverse even than crows when it comes to matters of taste. At least the birds only peck around in the faecal matter of others and restrict themselves to eating the seeds they find there, but rabbits and hares devour their own excrement. Though it must be said that they don't eat all their droppings indiscriminately, just the special ones. Like all herbivores, they have bacteria in their gut that help them dissolve and digest chewed grasses and other herbaceous matter. There are specialised species in the caecum, in particular, that break down greens into their component parts. However, some of the end-products can only be absorbed in the small intestine, which, annoyingly enough, comes before the caecum. And so the beneficial brew slides unused through the digestive tract and inevitably ends up in the outside world once again. What better idea than to smack your lips and enjoy this product from the caecum fresh from the anus, ingest it anew and extract valuable calories during its journey through the small intestine? The final processed waste – hard droppings – are completely ignored and are clearly regarded as faecal matter.[62]

We cannot imagine eating excrement, whether it's from animals or our own. Or at least most of us can't. And yet there are people who do this. And you can find them in Central Europe. They are hunters. To this day they shoot snipe, a hunt I find as repulsive as the hunt for whales. To make it even worse, these birds have hardly any meat on them, and perhaps this is why people have developed an odd way of eating them. The parts of the bird they enjoy include 'snipe mess', which is the gut along with its contents (that is, faecal matter). Finely chopped and dressed up with other ingredients, perhaps with bacon, eggs and onions, the whole concoction is baked on bread – and there you have it: a hunter's delicacy. Despite the fact that cooking kills worm eggs and other things you find in bird faeces, just thinking of such 'treats' makes me completely lose my appetite.

Animals must be able to taste things in order to tell the difference between what is appropriate as food and what is not – and also what is poisonous. By all accounts, however, it seems that many other species don't taste things the way we do. For example, there's a reason the popular children's book character Pooh Bear likes honey and his friend Tigger doesn't. Large carnivorous cats such as lions and tigers, and even marine carnivores such as sea lions, have lost their taste receptors for sweetness over the course of evolution. Clearly, sugary food is of no particular interest for these animals, and it's easy to understand why: meat doesn't taste sweet at all.[63]

It is even more difficult to compare our sense of taste with that of butterflies. Consider the swallowtail. The female lays her eggs only where her young have immediate access to the juicy leaves of suitable plants. That way, all the freshly hatched caterpillars have to do to satisfy their hunger is munch on the leaves around them. However, the butterfly doesn't have to use her mouthparts to test each plant on her way to lay her eggs, because she can check them out with her feet. As she walks around a leaf, her feet, which are equipped with sensory hairs, can taste up to six different substances. And that is not all. The butterfly can even detect the plants' age and health.[64] This may sound unbelievable, but we too can detect whether something is fresh or past its prime just by tasting it; think of an overripe banana, for example. And being able to taste a plant's health could be crucial to the survival of a butterfly's young. After all, if a plant dies before her caterpillars pupate, their dreams of metamorphosing into butterflies are over.

Something Special in the Air

HAVING EXAMINED TASTE, it seems obvious to take a closer look at the sense of smell. Animals definitely have a sense for what smells good and what smells bad. This sense serves not only – as with taste – to test food, but is useful in other ways that are also familiar to us. One of these is making oneself attractive to the opposite sex. I have to say that in autumn our billy goat, Vito, demonstrates just how far humans and animals diverge when it comes to the scents we experience as pleasant. As I mentioned earlier, he uses his own perfume – his urine – to make himself attractive to the two female goats in the herd. Because of this, my wife changes her clothes and wears a cap when she visits them in the stable, because the penetrating smell doesn't only drift through the whole yard, but also infiltrates fabric and hair.

But what we find disgusting might just be due to the culture of our times. Acquired tastes probably explain why deodorants were not yet in common use 200 years ago. Out on a military campaign, Napoleon supposedly once wrote to Josephine: 'I return to Paris tomorrow evening. Don't wash!' The sixteenth-century Spanish conquistadors, wanting to differentiate themselves from the fastidiously clean Moors whom they had just driven from the Iberian Peninsula, also had their doubts about the benefits of bathing. And when the Aztecs in Mexico met these light-skinned strangers for the first time, they immediately smelled the difference between them and their own people, who used steam baths for their personal hygiene, and thought it was nauseating. Old, well-aged cheese is a contemporary olfactory example. Dried-out milk protein giving off fumes that you could also describe as rotten and which, under other circumstances, would make people gag. I list these examples not to put humans on the same odiferous level as animals that stink. Not at all. I am merely pointing out that people, too, differ greatly in what they perceive as smelly.

Dogs trump billy goats when it comes to stink. Our dog Maxi loved to roll around in fox faeces, which have a particularly penetrating aroma. Fresh cow pats were another source she liked to use for special scents. For a long time people assumed that their four-legged companions rolled in dung to disguise their own scent. The theory was that this gave dogs – or at least their wild ancestors – better luck when hunting. Today, dogs and wolves are thought to use scent to relay messages – or perhaps they simply want to stand out in the pack. And clearly they don't find the smell, especially the stench of rotting meat or herbivore faeces, unpleasant. Quite the opposite.[65] Doesn't that remind you of people and their relationship to perfume?

You just need to be careful if your dog rolls in, or even eats, fox or dog faeces. Fox faeces in particular may contain fox tapeworm eggs. These eggs are as tiny as specks of dust, and after such a faecal bath they shower down from the fur of your beloved pet – likely as not, all over the living room. You then take the place of the mouse that was the eggs' intended host. The developing larvae settle in internal organs and make the mouse sick, to slow it down. Foxes then catch the slow-moving mice, which completes the worms' life cycle. But not, of course, if you are acting as the intermediary host instead. People get a severe infection that can be difficult to cure, depending on how far the worm infestation has spread. When your dog comes home decorated with fresh faeces, this is one time when it's definitely a good idea to give it a thorough drenching in the shower.

Even though their rating system differs from ours, animals are aware not only of things that smell good, but also of things that stink. Particularly when it comes to their own excrement. Herbivores are better off not browsing in places where they've deposited a pile of droppings, because practically every deer, goat or cow has worms, and a correspondingly large parasitic legacy can be found in faeces. Take lungworms, for example. For every 30 grams of faeces, there can be up to 20,000 eggs, which are then re-ingested when the animals browse. Because massive worm infestations physically weaken the host, infested herbivores are the first to fall prey to lynx and wolves. So it is only logical that an animal perceives its own excrement as absolutely revolting, to ensure it gives it a wide berth.[66]

I believe that the smell of their own faeces is just as awful for most animals as that of human faeces is for us.

Many domestic animals indicate that this is indeed the case. In the pasture, our horses retreat to a private corner that they use only for defecation. Out in the wild, thanks to horses' ability to move around freely, there would be no danger of eating in the same place too often. When we inhibit such movement, horses make do by finding appropriate spots in the pasture that they use only for defecation. Even Blacky, Hazel, Emma and Oskar, our rabbits, designate a toilet area in their hutch and run, where they do their main business. The only place where this doesn't happen is on factory farms. In these conditions, chickens and pigs even have to bed down in their faeces. The only way to avoid severe worm infestations on these farms is by regular doses of medication. It's a shame the pills don't get rid of the stink at the same time.

Many animals are as embarrassed about defecation as we are. When Crusty, the bulldog, was on a leash, he would pull away from us into the undergrowth if he had major business to attend to. What's more, he would turn his back to us, so that he didn't have to look at us. Clearly he was embarrassed to be seen when he was squatting. Apart from the smell, it is also just as important for all animals to be clean. Like us, they feel very uncomfortable if they have faeces or other dirt stuck to them. It's probably the reaction of their fellow animals that reinforces their feeling of discomfort. An animal with a dirty backside signals that it's probably ill and that's why it has diarrhoea. Other animals don't want to catch something nasty and certainly don't want to mate with a soiled partner.

And that's why, like us, animals take pains to clean themselves. Although it must be said that 'clean' means something different to animals than it does to us. Wild boar, for instance, love to cool down in summer, and they do this by enjoying a good wallow in a mud hole. Grunting and twitching their tails, they take their time rootling around, stirring up the mud and then settling down for a wallow. By the time the procedure is over, the pig is covered with a layer of beige-coloured mud. And yet the pigs don't feel dirty. And why should they? Isn't the human equivalent an expensive spa treatment with a pack of mud or peat? The wild boar feel the same – all fresh and clean. And there's a reason for that. As the mud coating dries, many parasites such as ticks and fleas are baked into it. Once the crust has hardened, the pigs rub it off on specially selected rubbing trees. They use the same trees or stumps for many years and, over time, they are rubbed smooth. Afterwards the pigs are free not only of the little animals plaguing them, but also of old bristles and hairs, which can be just as itchy. It's the same for our horses. They, too, like to roll around on the ground, especially when they're losing one season's coat and replacing it with the next. Depending on the weather, they can also end up with mud all over them – just mud, mind you, not faeces.

Comfort

THE LANDSCAPE IN CENTRAL EUROPE, as in many other
places in the world, is a strange patchwork, at least
when seen from the point of view of a wild animal. Wide
open areas unbroken by roads or settlements are a thing
of the past, and you couldn't lose yourself in the wilder-
ness any more, even if you tried. For even our most
natural ecosystems – our woodlands – are not what they
used to be. In Germany there are now 13 kilometres of
logging roads per square kilometre of wood, so that
logging trucks can reach the remotest areas. Chances
are that on a cross-country trek you won't get farther
than 100 metres before coming upon a trail. This means
that your adventure will consist primarily in choosing
the wrong fork in the road.

Trails have distinct disadvantages for the natural
world. The once-loose soil is hugely compacted, and the
tiny creatures that used to live in its subterranean layers
have all suffocated. In addition, trails act like dams to
block the flow of water, and this interference should not
be underestimated. Numerous streams flow underground, and in many cases they are blocked or diverted
by the compacted soil under trails. This means some
sections of woodland are transformed into parcels of
swampland on which many trees fail to thrive because

their roots suffocate in the foul brew. Woodland trails also pose substantial obstacles to light-averse ground beetles. The beetles, which lost the ability to fly long ago, don't dare leave the darkness under the trees to cross paths flooded with light. Confined to a small territory surrounded by trails, they can no longer exchange genetic material with their neighbours.

But trails don't have to be a hindrance to animals. Like us, deer and wild boar walk, and they prefer not to struggle through the undergrowth. A stroll through wet grass or bushes in rainy weather is unpleasant, which is why four-legged creatures are happy to use our nice, smooth cross-country game routes. For that's how game see our roads and trails: as game routes provided by people. The animals find it much easier to walk along them, as you can see from the numerous tracks pressed into soft spots on the surface. And where people aren't there to help out, the animals make these cross-country routes for themselves. They are, mind you, considerably narrower, measuring only the width of the animal. There's no ordered plan. One day the lead sow in a herd of wild boar finds her way through the underbrush. The other boar follow, and the grass and other greenery are soon trampled down. Next time the pigs pass there are still traces of this faint track, and it's a little easier to walk along it. Over the course of time and after years of use, the path looks like a footpath trampled down by people: all the vegetation has been crushed by trotters and the path shows up as a narrow strip of bare earth. Knowledge of these easily walkable cross-country routes is passed down from generation to generation, unless humans come along and obstruct them.

One day early in my work managing the forest, I had a fence built around a new planting of oak trees. There were way too many deer, and they would have been only too happy to eat the seedlings' juicy shoots, and so I had to protect the tiny trees. Later on, I figured out that the fence had interrupted a well-established long-distance route used by these large herbivores and had forced them to find ways around it. This led to many dangerous incidents with cars, because deer began popping up in unexpected places. The fence has since been removed, and the deer have returned to using their traditional path.

Incidentally, people create trails the same way animals do. I observed the process in the part of the forest I manage that has been set aside for burials – what we call our 'final forest resting place'. This is where we lease ancient beeches to serve as living grave-markers, and funerals are held when urns are buried there. The burial ground saves the 1,000-year-old woodland from the axe. To keep the area as natural as possible, the forest service decided not to create any new roads or paths. Nonetheless, a few beaten paths have appeared in places where it is especially easy to walk through the trees and their million-strong offspring. Rain helps form these paths. When a front of bad weather passes over the wood, the leaves of young beeches drip with moisture for a long while afterwards. No one wants to brush past the leaves because their trouser legs would get soaked in seconds. And so people seek out routes where they can stay at least partly dry, and those coming later do the same thing, following these faint traces of a path. That's fine with me, because it means the impact of visitors is confined to a small fraction of the forest floor.

I have to add that the animals' cross-country routes also offer disadvantages to those using them, because the heavy foot traffic attracts uninvited guests. More prevalent than the predators that lurk along the paths waiting to make a meal of careless travellers are small critters waiting for their next meal: ticks. Ticks are members of the mite family, and they depend on meals of blood. Because they move very slowly, they have to wait for their victims to come to them. And where better to wait than along a well-travelled path? Here, ticks cling to stalks of grass, twigs or leaves hanging within range of the backs of deer or wild boar, waiting for the scent of mammalian breath or sweat and the vibrations caused by approaching hooves or trotters. So if you're wandering through the woods in summer, it's best to avoid cross-country routes frequented by game. In winter, it's not a problem, because ticks are not active at low temperatures. And there are more stowaways than just ticks lying in wait along these game routes. Plants, for example, are also waiting for a ride – specifically a ride for their young. That's why the tiny fruit of catchweed are equipped with barbed hooks. When an animal passes by and brushes up against the plant, it takes a portion of the catchweed's seeds along with it, which will fall off farther along the path. Researchers have shown that plants like catchweed spread specifically along paths frequented by game.

But let's come back to getting your legs wet. You might have experienced how unpleasant this is, when you are out walking. Why should it feel any different for animals? They feel freezing cold when their coats are wet, and they would prefer to stick to the easy trails. And these paths offer one further advantage: speed of travel. When branches are snapping in the bushes and an enemy is preparing to pounce on a fawn or a piglet to eat it, the herd wants to be able to run away as quickly as it can. And because thick branches and downed trees turn flight in the forest into a scramble over an obstacle course, the best place to run is along an open trail.

Weathering the Storm

FEW OF US WOULD VOLUNTARILY enter a forest in a storm. Lightning strikes in trees can be deadly, and pelting cold rain isn't pleasant, either. For a number of years I offered survival training in the woodland I manage, where the participants spent a weekend in the woods equipped with nothing but a sleeping bag, a cup and a knife. They slept there and, most importantly, foraged for food. One time when we were foraging, we were surprised by a violent storm. We had no choice but to let it beat down onto us. Apart from the soaking rain, lightning strikes nearby were a cause for concern. I bluffed my way through the storm with exaggerated calm, so as not to worry my companions even more. Inside, however, I felt some rising panic when a powerful strike hit only about 100 metres away. I've often noticed after a thunderstorm that even if you are not hit directly by lightning, the area around a tree that has been struck is also a dangerous place. On one occasion not only did the tree that was ripped open die, but so did more than ten other close neighbours. In one extreme case, I might as well have been watching a knife-throwing act. The lightning strike caused so much pressure in a spruce that its wood splintered into multiple shards that flew through the air with such force that some of them impaled a nearby stump. That day of survival training, however, we were rewarded with a wonderful wildlife sighting in the calm after the storm. The rain stopped abruptly and a break appeared in the clouds, allowing the sun to shine through, bright and warm. The vegetation around us started steaming, and suddenly a deer shot out into a small clearing. The animal was completely drenched and was looking for a warm spot to dry out. He felt just the same way we did, and I felt an instant moment of connection.

What's it like to be a wild animal anyway? They have to tough it out year-round in the wind and the weather, and when the cold season rolls around, that can't be very pleasant. Or can it? Let's take a closer look. First of all, animal fur repels much more moisture than is commonly thought, thanks to a coating of oil – the one we're constantly washing out when we reach for the shampoo. Then, the direction in which the hair grows on an animal's back points the ends down, so the individual hairs act like roof tiles and channel water towards the ground. Thus the coating of hair on deer and wild boar keeps the rain off, and their skin stays nice and dry as long as a strong wind doesn't blow the rain sideways and up between the individual hairs. Older animals are well aware of this and, when forced by the weather, they change location to somewhere well protected from the wind. Moreover, they position themselves so that their rear end is pointing towards the wind, and their head, which is more sensitive, is sheltered from it. The only time they experience problems is when snow falls and the temperature hovers around freezing, because then the snowflakes melt and form rivulets that run down between the hairs, making the animals shiver. They prefer a hard freeze. Their thick winter coat insulates them so well that, when it's nice and cold, a layer of freshly fallen snow can lie on their backs for hours without melting.

Don't we feel the same way? Aren't we happier on a clear, frosty day at minus 10 degrees Celsius than in windy, rainy weather at 5 degrees? Animals don't really feel any differently: it's just that in general they endure lower temperatures better than we do. But even that is not absolutely true, and to illustrate this I have to mention another survival training weekend. Years ago I once held the training in winter, and the weather on that January weekend was particularly awful. The temperature hovered around freezing, and every hour rain turned to snow and back again. Even the firewood was so wet that it was difficult to keep the camp fire burning. I was expecting that it wouldn't take long for the participants to give up, but after a night in cold, clammy sleeping bags, our bodies seemed to adapt and none of us felt cold any longer; apparently we had reached the comfort level of wild animals.

In summer, apart from the warmth of the sun, there's yet another reason for animals to abandon the thick canopy of leaves under the trees for a small clearing after a downpour. The leaves of beeches and oaks keep dripping for so long after a rainstorm that there's a local saying: 'In deciduous forests, it always rains twice.'

But the fact that deer get rained on even after the storm has passed is not the only thing that disturbs them: falling water droplets are loud. Prey animals can't hear approaching predators behind this curtain of sound, and the predators like to take advantage of this to do a little stalking. And that is why deer prefer to move to a clearing right after a cloudburst – so that they can listen to see if everything is still as it should be.

The situation for small mammals such as voles is more challenging. Sometimes when I walk through our horse pasture in winter rains, I can see water bubbling up out of the vole holes on the hillside. How are the tiny rodents supposed to survive such conditions? Wet fur is much more dangerous for them than it is for large animals, because percentage-wise for their body size they lose much more warmth. Besides, they have, comparatively speaking, an enormous need for calories: every day they must consume the equivalent of their body weight in food. When they get wet, their energy needs to increase considerably. And because they don't hibernate, they don't get any break from the daily struggle of finding food. On the plus side, they prefer to eat the roots of grasses and other green stuff, and therefore they don't need to brave icy winter winds, but can gather food from the comfort of their underground tunnels.

But what happens when water inundates their earthworks? The smart little animals have a special building plan to deal with this eventuality. For starters, there's a chute at the entrance. In case of danger, a vole can drop into this to flee quickly underground. Then the tunnels go deep, far deeper than is strictly necessary. After a short distance, they angle slightly upwards, ending in a small chamber comfortably padded with soft grass. If it rains so hard that water runs into the burrows, the water collects in the deeper sections of the tunnels while the inhabitants sit comfortably high and dry. And because a multitude of tunnels weave their way through the burrows, the voles can flee if water does manage to reach their nests. However, the plan doesn't always work. When the whole pasture is under water after a heavy rain, which happens mostly in winter, some of the voles inevitably get caught and drown miserably in their underground chambers.

Pain

——

IT WAS A COLD EVENING in February and our goat Bärli was about to give birth. She was restless, she kept lying down and milk was already flowing into her udder. My wife was worried. 'I think it's taking too long,' she insisted. 'Shouldn't we call the vet to be on the safe side?'

I calmed her down. 'Bärli can do this on her own. Perhaps she just needs some peace and quiet. She's strong and healthy. I don't want to interfere unnecessarily in a situation like this.'

Well, it would have been better if I had paid attention to Miriam and her sixth sense. The next morning the kids were still not born and Bärli was clearly in pain. She was grinding her teeth, she didn't want to get up and she didn't want to eat, either. All these signs were highly alarming, and we lost no time in calling a vet who knew our goats well. 'He's on holiday,' the vet who was covering for him told us, but she quickly drove out to our forest lodge. She diagnosed a breech that unfortunately had already died inside the mother. The vet carefully pulled out the dead baby and then gave Bärli medicine to get her to expel the placenta.

Our goat recovered quickly, and we even got her another kid to adopt. A goat farm in the vicinity had a quadruplet to give away. Four kids are too much for any mother goat. She has only two teats on her udder and not enough milk for so many mouths, and the farmer was happy to find a good home for one of the feisty little kids. We rubbed mucus from the dead kid over the new arrival (who was later to become our breeding billy goat, Vito). That might sound disgusting, but it meant that as far as Bärli was concerned, the new kid smelled like her own child and she immediately allowed it to nurse. Both mother and child thrived – so there was a happy ending for this pair, at least.

But back to pain. Pain? We have heard earlier about fish feeling pain, but those reports are still considered controversial. We could get down to the neurological level and advance all kinds of arguments about similar triggers and signals and patterns of brainwaves and hormones suggesting similar feelings. But isn't there a much easier way? After all, Bärli exhibited all the patterns of behaviour you see in people who are in pain. Grinding her teeth (something goats don't usually do), loss of appetite, lying down, apathy. Don't some of these symptoms sound familiar?

We've experienced more direct evidence in our dealings with our chickens, goats and horses. We keep all our animals behind electric fences appropriate for each, to make sure they stay in the areas we've marked out for them. An electric fence might sound cruel, but it's the most practical solution. Barbed wire is out, because of the chance of injury. And a wooden fence wouldn't keep the goats in for long, and the horses would gnaw away at the posts and bars over time. Just how an electric fence works is something I experience myself on a regular basis. Sometimes when I go out in the morning, lost in thought, to mark off a new section of the pasture for the horses, I forget to switch off the power first. Then a powerful shock rudely brings me back to earth, and I get annoyed with myself. For the next few days I check multiple times to see if the pasture fence control really is shut off. An electric shock is a wake-up call for the instincts, sending them into overdrive.

And that's exactly how the fence works for animals, too. They experience how unpleasant it is to touch the fence once, maybe twice, and from then on they avoid it. So an electric fence works by means of an initial shock and then by the mere memory of the shock. Just like it does with me. And that's why I'm convinced our domestic animals experience the electric shock exactly the same way I do. And not just our domestic animals. With the chickens, the primary purpose of the electric net fencing we use is to keep foxes out, and it works very well. Farmers fence their maize fields with electrified wires to keep wild boar out. And pet owners who don't like visible fences can bury cables. When a dog or cat crosses the invisible boundary, they receive an electric shock from a special collar. It's up to the individual to decide whether that's okay or not, but the fact remains that all these creatures feel pain and instinctively react the same way – including me.

Fear

—

A PERSON OR AN ANIMAL that doesn't know fear will not survive, because fear protects us from making deadly mistakes. Perhaps you are familiar with that queasy feeling you get when you're up somewhere high, like a viewing platform or the Eiffel Tower in Paris. I get a tingling sensation and the overwhelming desire to climb back down again as quickly as possible. That makes a great deal of sense, from an evolutionary perspective. This inborn instinct prevented our forefathers from falling off high cliffs, which would have abruptly severed the generational line that leads to us.

Wild boar show us that animals not only experience the gut-wrenching feeling of fear and recognise threats, but can also process this information and use it for long-term planning. So let's take a little excursion to Switzerland, and specifically to the canton of Geneva. In a public referendum in 1974 the people of Geneva voted in favour of a ban on hunting. Hunters are large mammals' worst enemy. And because hunters belong to the species *Homo sapiens*, animals that are hunted fear humans. That's why they prefer to spend the day in thick forests and brushy areas, out of the sight of their dangerous two-legged adversaries, and usually venture out into pastures and fields only at night. When hunting was outlawed in Geneva, the behaviour of deer and wild boar there changed. They lost their wariness of people and now show themselves during the day. But it's not only the Genevese wild boar that have changed their behaviour. All around, including in neighbouring France, hunters are still blasting away. And so as soon as the hunting season kicks off – especially in the autumn, when the drive-hunts that use packs of hounds start – the pigs become excellent swimmers. As blasts of the hunting horn reverberate through the air and the gunshots begin to ring out, many boar leave the French shore and swim across the Rhône river to the canton of Geneva. Safe on the other side, they can now thumb their snouts at the French hunters.

The swimming swine show three things. First, they are aware of the danger and can remember the previous year's hunt, when family members brought down in a hail of bullets had to be left behind dead or severely wounded. Second, they must be afraid, for this is what drives them to abandon the territory where they have felt so comfortable all summer long. Third, they must be able to remember that they will be safe across the water in Geneva. Over the extended period of more than four decades, the journey has become a tradition passed from one generation of wild boar to the next: in times of danger, we cross the river to safety. The forefathers of these omnivores discovered this through trial and error in the 1970s. They obviously have a well-developed sense of self-preservation.

We know, from the example with the electric fence, that animals can become fearful just by remembering something. You know yourself how particular songs, smells or pictures can dredge up memories of threatening experiences from the depths of your unconscious brain. Well, the same mechanism works with dogs. If you have a canine four-legged family member, you might have had a similar experience to ours. Our little Münsterländer, Maxi, loved life and a change of scene; she just didn't like the vet. When she went there, she had to endure needles and sometimes there was unpleasant tartar-removal or the somewhat gross squeezing of her anal glands. Little wonder that every time she went, Maxi stood trembling on the examining table, a little heap of misery as she submitted to each procedure. But not only that. While she was still on her way to the vet, the little dog registered the telltale smells of the neighbourhood through the car's ventilation system and she began to be afraid the moment we turned into the parking lot. A specific film must have been playing through her head that anticipated the unpleasant situation. So let's grant that there is good evidence to show that animals can feel fear. But our little dog's reactions point to something else as well. Given that there was sometimes a gap of more than a year between visits to the vet, it is clear that dogs – and other animals – can remember things for a very long time (just as our goats remember the electric fence).

However unpleasant it sounds (and is), most wild animals are like Maxi. As soon as they spot us, they get anxious, especially when we're too close. Apart from fear, it would be interesting to know what other reactions we trigger in them. Can they tell us apart from other animals? Do they have any inkling that we build computers, drive cars and tower above them intellectually, at least in some respects? Think about it from our point of view. Apart from our pets, no single species of animal has a particular significance for us, none stands out from the others. So does that mean it is all the same to a deer whether it sees a human, a hawk or a hedgehog? Theoretically, yes. You can test this out for yourself when you think back to the last time you walked in the forest. Perhaps you remember rare or particularly large or colourful species, but can you remember every bird or describe every fly? You can't, because we take it for granted that our surroundings are full of life, and so we don't bother to notice details about all the creatures crawling and flying around us.

It's difficult to get a clear idea of how other people see the world, because it's almost impossible to get inside their heads. If we can't even do that, how can we hope to see things from the perspective of animals? The easiest way to try and do this is by looking at animals' reactions when we appear. What's important here is whether or not we play a significant role in their daily lives. On the one hand, we could have an impact because of the pain or even death we inflict while exploiting or hunting them; on the other hand, we could have an impact because of positive aspects of caring for them, such as providing them with food. The situations I find particularly gratifying are those where we are not involved at all – when we are neither harming nor helping them. In these circumstances, animals usually behave the way they do in representations of paradise: they basically ignore us. A particularly extreme example all the way from Africa was shared on the Internet in summer 2015. The online edition of the *Daily Mail* posted photographs from Kruger National Park in South Africa. Lions were standing in the middle of a busy road tearing the flesh off an antelope, with cars all around them. What surprised and shocked drivers most was that the predators were completely uninterested in the backdrop to their meal: bushes, stones, people in cars – it was all the same to them.[67]

More harmless experiences of animals ignoring people are the photo safaris you can take in Africa where you can park close to zebra, wild dogs and antelope. On the Galapagos Islands, on the coast of Antarctica, in the marinas of California and around Yellowstone – all over the world – you can find places where animals allow people to get very close, without getting wary. So why isn't that the case where I live in Central Europe? After all it has one of the highest concentrations of mammals in the world. There are about fifty roe deer, red deer and wild boar per square kilometre of wood. And although theoretically you should be able to see these animals at any time of the day or night, you usually come across them only at night. You've already learned the reason for this: in Central Europe these animals are hunted wherever they appear.

People are visual animals, and we hunt by sight. And so the goal of our potential prey has to be to disappear from view. If we hunted by scent, then perhaps over generations animals would have become almost odourless. If we hunted by hearing, they would perhaps have become extremely quiet. Instead, they endeavour to remove themselves from our field of vision. The most important element here is the time of day. It's quite simple: because we see barely anything in the dark, our prey evade us by being active at night. We think deer and wild boar are naturally nocturnal, but they're not. They need food at regular intervals around the clock. But instead of spending their days in meadows or along the edges of woodlands, which would be their normal behaviour, they spend their days feeding out of sight in the underbrush or deep woods. They feel confident enough to leave their hiding places only when darkness falls – the time when people are optically challenged. It is only the very hungry or careless youngsters who venture out earlier and dare to enter areas where hunting blinds loom over the landscape. We euphemistically call these 'raised hides', but to deer they are death-traps where their worst enemies perch and deal sudden death with a bang and a puff of smoke.

That's not just my view: it's quite clear to foresters and hunters alike that wild animals learn from experience. This is how a herd of deer experiences the shooting of one of their own. There's a crack and a sudden smell of blood. The shot is often not a clean one, and the wounded animal manages a few metres of panicked flight before its legs buckle and it falls. This sight, coupled with the smell of stress hormones, imprints itself deep into the consciousness of the other members of the herd. And when there's a creaking and crashing from the hide, as the hunter climbs down to salvage the downed game, the intelligent animals make the connection. From then on, before stepping out onto the path, they look suspiciously up at the raised hide to see if someone is sitting there. They could, of course, give the area a wide berth, but hunters usually set up their hides in places where particularly tasty food grows. And if there isn't any, they simply sow an appropriate mix of attractive meadow plants. These mixes have names like 'Gameland Goulash'. Doesn't that sound enticing? And so the evenings become a never-ending game of roulette. If hunger wins, then the deer step out onto the path too early and therefore into the shooters' visual range. If fear wins, the hungry deer don't come out to the feast laid out for them until it's pitch dark, and the hunters go home empty-handed.

Foresters in Eifel National Park have observed just how sensitive deer are to their surroundings. A forester who hunts in the area and a forest-service employee who works there owned the same make of car. Although the game beat a retreat as soon as the forester's car appeared, they stayed put when the forest-service employee drove down the road. But it's not just deer that have mastered the skill of telling the difference between dangerous and harmless people. Our domestic animals also rely on their gut instinct here: deer and

company want to avoid hunters, whereas – as we've just
seen with Maxi – dogs and cats want to avoid the vet.
Hunters are considerably more dangerous than vets, and
so it's no wonder that quite a few species of animals
can tell what kind of people are out and about. Take
the European jay. Whereas almost all animals consider
children to be harmless, jays tend to tolerate adult hikers
as well. That is, until hunters are on the loose. Then
jays kick up an enormous racket and warn the animals
nearby with their raucous calls. This is why many
hunters continue to target these colourful birds, which
is unfortunate because jays play an almost indispensable
role in distributing tree seeds in the woods.

   Wild game animals are getting stressed when
people come into their territory. If two-legged intruders
are constantly entering the area, the portion of time an
animal needs to spend checking to see if it is safe rises
from 5 per cent to 30 per cent of the day.[68] At least that
holds true for those people whom animals find it difficult
to categorise. Walkers, cyclists or horseback riders who
keep to the trails are easy to evaluate: they make noise
and move along clearly demarcated routes. As long as
the game can see them, it's clear they are moving directly
from A to B, and animals observing them from the safety
of their daytime hiding places have nothing to worry
about. In contrast, mushroom-hunters, mountain-bike
riders, hunters and foresters often travel off the beaten
track. And because most of these individuals travel alone,
there's no lively conversation that wild game can listen
in on and use to estimate the route the intruders are

taking. All the game can hear is the occasional snapping of a twig under a boot and perhaps a quiet clearing of the throat from time to time – nothing more. Then animals become uneasy and, to be on the safe side, they prefer to move away quickly.

You might argue that things have always been this way. What difference does it make whether it's a wolf pack or a person on the hunt? Well, one big difference is the number of hunters. Contrast one four-legged hunter for every 50 square kilometres in wolf territory with about 10,000 two-legged predators tripping up over each other in the same amount of space in Germany today. And the wild game can't tell right away that not everyone is armed. So if there's any doubt, game retreats from each potential attacker and generally forgoes forays into lush meadows in the bright light of day. As you can see, life for game that can be legally hunted is quite nerve-racking. Their situation, where for every potential prey there are any number of potential hunters, is unique in the animal world, because the situation is usually the other way round.

Thus it's hardly surprising that fear and mistrust are widespread out in the woods and meadows. Let's take a look at which animals have to endure the stress of being hunted. I've already mentioned red deer, roe deer and wild boar. As far as mammals go, you can add to this list chamois, mouflon sheep, foxes, badgers, hares, martens and weasels. There are also quite a few species of birds, such as partridges, various pigeons, geese, ducks, gulls, snipe, grey herons, cormorants and corvids.

Is it any surprise that we hardly ever get to see any of the birds in this colourful array? Imagine if the situation were reversed and there were between 2,000 and 3,000 lions wandering through each square kilometre in Central Europe. That roughly corresponds to the numerical dominance of two-legged predators that confronts hunted game in this area. Now let's get back to the animals' point of view. Here – I don't know about you – but my powers of imagination at least have reached their limit. If there was deadly danger lurking in every corner and behind every bush, I wouldn't dare leave my house. If I had to, I would come out only at night, when I knew that my persecutors would definitely be sleeping or, if they were up and about, at least they wouldn't be hunting me down.

An animal that has seen a family member collapse covered in blood, or has experienced fear and rising panic deep in its bones, will pass these experiences on, probably over many generations. Researchers have concluded that this transfer happens even in the absence of language, for – as reported back in 2009 – fear is not only felt in the bones, it is also expressed in the genes.[69] The Max Planck Institute of Psychiatry in Munich discovered that during traumatic experiences particular chemical markers (methyl groups) get attached to genes. They work like switches and alter the activity of the genes.[70] According to the discoveries researchers made using mice, this means that behaviour can be changed for life. The research also predicts that, thanks to these altered genes, certain patterns of behaviour can be inherited. In other words, our genetic code passes down not only physical characteristics but also, to a certain extent, experiences. And what experience could be more traumatic than the severe wounding or death of your next of kin? It is not a pleasant thought that the majority of the animals living around us are traumatised.

Thankfully, the cohabitation of wild animals and people also has its pleasant side. There is hope that even in Central Europe we can live peaceably with one another, as we can see from the increasing number of wild animals living in urban areas. The news is going round in animal circles that there's a kind of sanctuary set up in cities. And, indeed, built-up areas do count among the so-called no-go areas where hunting is banned. From the animals' point of view, the only difference between cities such as Berlin, Munich or Hamburg and national parks is that the land has been developed. And what animals do you find in these urban centres? Wild boar in the front garden happily rooting around in the tulip beds, ignoring attempts to drive them away (why would they leave?); foxes digging their dens in roadside embankments; raccoons setting up house in garages and attics. The animals are making themselves at home right in the middle of the spaces we've tamed. Whereas, to us, tarmac and rows of drab houses symbolise a disconnect with Nature, what animals see is an environment extraordinarily rich in vertical rock faces, where all the mountain tops are bizarrely cube-shaped. Urban areas are increasingly revealing themselves to be ecological jewels. Berlin has one of the highest populations of goshawks in the world – about 100 breeding pairs.[71] The birds nest in the city's parks, using them as a base from which to hunt rabbits and pigeons. I myself have seen a fox near the Brandenburg Gate calmly munching a discarded curry sausage.

Not every city-dweller can deal with this proximity to Nature. An older woman explained to me that she was frightened a fox might appear at her patio door. Thoughts of rabies or fox tapeworm flashed through her mind, spoiling the anticipation of what might turn out to be a wonderful wildlife encounter. In fact wild animals in Central Europe present few dangers. Rabies was eradicated many years ago, and fox tapeworm is quite rare, at least in the wild. I've already mentioned how the chain of infection passes from mouse to fox, and the problems with fox faeces. If a dog eats an infected mouse (and there are lots of dogs that hunt mice), then it too excretes thousands of eggs when it does its business. Moreover, after it licks its rear clean, it licks its fur, meaning that it can distribute tapeworm eggs no bigger than specks of dust throughout the house. A dog that is not regularly dewormed poses more danger than a passing fox.

Perhaps we exaggerate dangers from the wild simply because if we didn't, there wouldn't be much left to fear. Do our archaic instincts simply need something 'dangerous' to react to? Granted, it's a bit different with wild boar when they have piglets. A friend of mine who lives in Berlin-Dahlem told me the animals won't leave his garden even when he tries to scare them away by clapping loudly – there's not much more a city-dweller can do.

The kite, a large bird of prey, is another European species that seeks out proximity with people and has preferences about who to hang out with. In earlier times the birds were hunted and persecuted; however, since they've been protected, they like to spend time around people – that is, they like to spend time around people who own tractors. When the fields are cut in summer, the kites profit from farmers' labours. For the heavy machines don't only cut grass, they also dispatch numerous mice and other small animals into the great beyond. That doesn't sound nice and it isn't good; however, for the kite, this is literally found food. In Hümmel you can spot the majestic birds as soon as the tractors drive into the fields to begin work. With a wingspan of just over 1.5 metres, they glide low to the ground, following the machines and keeping an eye out for flattened mice or ground-up fawns.

Martens are a less welcome sight, even though they are really beautiful to look at. They are not hunted in urban areas, and trapping, which used to be common in woods and fields, has declined considerably, so martens have mostly lost their fear of people. We once raised an orphaned animal that allowed us to stroke it and made a kind of contented purring sound, just like a cat when it feels good. At first we fed the youngster from a can, but then to prepare it for life in the wild we also gave it mice for breakfast. It didn't take long for the little chap to become so wild that we could only hold it when we were wearing gloves. Finally, we opened the door to its cage so it could decide for itself when to leave us. It took three nights, and then the cage was empty and we never saw the marten again. But perhaps it's still scampering around somewhere on our property in the dark, although if it were, it would be more than ten years old by now.

It's questionable whether we did ourselves a favour when we helped the marten. There are two cars parked in front of our forest lodge: a four-wheel-drive vehicle for work in the woods and a car for private use. One day I noticed a piece of rubber hose lying on the ground near the bonnet of the jeep. I lifted the bonnet right away and found a nice mess. A marten had done a thorough job of biting through a number of cables and hoses. There was nothing for it but to take the vehicle into the garage for repair.

But what had induced the animal to go on a rampage in the engine compartment, and why is the marten sometimes gripped by such a destructive mania? By the way, there's no such thing as 'the' marten, for there are two species in Central Europe: the pine marten and the beech marten (also known as the stone marten or the white-breasted marten). The pine marten is a shy denizen of the woods that likes to sleep in hollow trees and spends the rest of its time scurrying through the branches up in the canopy. The beech marten, however, is not so attached to trees and feels at home in other places as well: rocky areas and caves or even houses, which, after all, are just square mountains. These are the places the curious beech marten looks for prey and, while it's doing so, it investigates everything with its sharp little teeth.

Severed cables, trashed hoses and ripped blanket insulators in the engine compartment, however, are not the result of curiosity, but of blind rage. And when these little beasts sense competition, they can certainly lose it. Martens use their scent glands to mark their territory, sending a clear message to every other marten of the same sex: 'Occupied – keep out.' Normally martens respect aromatic boundaries and leave each other alone. Because it's so nice and cosy under the bonnet of your vehicle, 'your' house marten makes regular visits to your car. Sometimes it drops off a few supplies: one day we found the lower leg of a rabbit on our car battery. However, these visits don't result in any damage. It's only when you park your car somewhere else overnight that things get serious. In the new location, other martens come on over and investigate the unfamiliar object, leaving scent trails as they rummage through the cavities.

Back at home, you've given your marten quite a puzzle. It has to assume that another marten has broken all the rules and entered its favourite hollow uninvited – the absolute nerve! It furiously tries to remove all traces and banish its rivals. Soft hoses are ideally suited for venting rage, and the marten doesn't nibble on them carefully, as it does when it's checking things out. In its blind fury, it brutally yanks them out. You can often see from the state of the blanket insulators underneath the bonnet just how incandescent with rage they are. Sometimes all you find are claw marks; however, when we looked under the bonnet of our old Opel Vectra, we found the material hanging down in shreds. To achieve this effect, the marten had clearly lain down on its back and thrashed around wildly, ripping out whole chunks of the mat with its sharp claws. The so-called 'car martens' don't necessarily love cars, they simply hate competition. If you park your car in the same place every night, you'll probably be fine.

Meanwhile, there are innumerable insider tips on how to scare the animals away. But small bags of human hair or sanitising toilet-rim blocks hanging in the engine compartment are examples of remedies that will only be effective for a few days at most. For a while we tried sprinkling pepper over the engine. That too didn't work in the long term, but a built-in device that delivered a shock when the marten stepped on electrified plates did the trick. We placed the plates where the marten usually got in and, after its first contact, it avoided them. An ultrasonic apparatus equipped with a bright light that reacted to movement was similarly effective. However, the animals become deaf to devices that broadcast ultrasonic sounds all the time, and the constant noise is detrimental to bats and other species, so I wouldn't recommend using them.

So what about our domestic animals? Do they idolise us and stay with us of their own free will? Or is it perhaps fear that keeps them close? If there's a fence involved, the question doesn't even need to be asked. Cows, horses and, yes, our goats as well are, strictly speaking, prisoners, even if they probably don't think of themselves that way. There's another unpleasant comparison here that has to do with Stockholm syndrome. The person who coined the term, criminologist and psychiatrist Nils Bejerot, was investigating the relationship between a man who robbed a Swedish bank in 1973 and his victims. The hostages developed feelings for the thirty-two-year-old kidnapper that were similar to those children have for their mothers, whereas they hated the police and the authorities. This paradoxical development has been diagnosed in many other situations and is considered to be a protective psychological reflex that helps the victims survive their dangerous situations and somewhat lessens the psychological damage they suffer.[72]

If animals are sensitive souls (and I think they are), perhaps they develop similar strategies. When they're held captive, they usually don't trust us right away: they are wary and keep their distance. It takes a while before they start calling out a friendly greeting when they spot us in the distance, walking towards the pasture. Does that sound somewhat heartless? Nature didn't intend for goats and horses to spend their whole lives as prisoners behind a fence. Let's not pretend: these animals would run away, if only they could. So if they do indeed develop some kind of Stockholm syndrome, it might actually be the best thing for them, because they will then accept their fate and find it tolerable.

When we work out in the pasture, we can tell that our goats and horses enjoy being around us. Of course the friendly greetings when we appear could also have something to do with feeding time, in which case we'd only be cheered as bearers of nourishment. It's a bit different with dogs and cats, although not initially. The relationship starts with an involuntary match-up. Then the animals are taken to their new home and are forced to spend a few days under house-arrest, or constrained at the end of a leash when out walking, until they get used to their owners. This means that acclimatisation is not completely of their choosing; however, when dogs and cats finally regain their freedom, they could just take off. But they don't.

More moving are the rare cases when an abandoned pet adopts a person. Completely unforced relationships like this demonstrate that true partnerships are possible. Moreover, these kinds of partnerships happen not only between people and animals, but also between animals of different species. Wolves and ravens are one example, as wolf researcher Elli Radinger explained to me. It turns out that ravens enjoy living with wolf packs, and the wolf pups even engage the black birds in play. When large enemies such as grizzly bears approach, the ravens warn their four-legged friends. The wolves repay the debt by allowing their feathered partners to feed alongside them at their kills.

High Society

HAVE YOU READ THE NOVEL *Watership Down*? It's a moving
story about some rabbits living in the English countryside that have to move out of their old home and look
out for a new one. When they find new territory, they
must fight the local rabbits until they finally conquer a
corner for themselves. We shelter a family of rabbits in
the garden at our forest lodge. Hazel, Emma, Blacky
and Oskar live in a small enclosure. They have weather-
proof accommodation and an open space where they are
free to roam. We get a good view of their social life.
There are fights and squabbles, but mostly there are
moments of tenderness. The rabbits lick each other's
fur or, on warm summer days, stretch out nestled up
against each other in the shade. Of course there is also
a hierarchy but, with only four animals, we can't find
out much about it.

Dr Dietrich von Holst, a professor at the University
of Bayreuth, has a completely different set-up. He created
a 22,000-square-metre research area for wild rabbits
and spent twenty years observing their behaviour.
Population levels were constantly changing. On the one
hand, diseases and predators wiped out up to 80 per
cent of the sexually mature adults; on the other, they
'bred like rabbits,' as the saying goes, so the group grew

to about 100 adults. Holst noticed that this population fluctuation didn't affect all levels of rabbit society equally. Rabbits live according to a strict hierarchy, which is different for each sex. Each rabbit vigorously defends its rank, and for good reason: dominant animals reproduce more successfully. Although the top males and females are more aggressive, overall they suffer less from stress. That makes good sense. After all, rabbits that are constantly being pushed around live in constant fear of the next attack. High-status rabbits experience elevated levels of stress hormones only in short bursts when they are attacking. No wonder Professor von Holst reported that the high-ranking rabbits experienced lower levels of stress.[73]

In addition, high-ranking rabbits have especially close social contact with rabbits of the opposite sex, which helps them relax. The average lifespan of an adult rabbit is two-and-a-half years, and there are clear discrepancies between rabbits at different levels of the hierarchy. Whereas rabbits on the lowest levels often die just a few weeks after reaching sexual maturity, rabbits of high social standing live for up to seven years. And that is not just because they get more to eat or are picked off less often by predators. No, the most decisive factor is lower stress levels. A life with less worry, and therefore more rest and relaxation, means less risk of intestinal illnesses, which are the leading cause of death in rabbits.

Good and Evil

ANIMALS ARE NOT NECESSARILY better-natured than people: they can be very aggressive – not only to other species, but to each other, as a quick look around our yard confirms. From the yard to the road there are four beehives, and the busy bees fly out into the surrounding landscape to gather nectar. That's exhausting work, because bees must visit 100,000 flowers to get enough nectar to produce 17 grams of honey.[74] They are not gathering their sweet cargo for me, the bee-keeper, however. The honey they make from it will serve as a source of energy for the bees, which have to shiver their way through the cold of winter. If things don't go according to plan over the summer and the stores of honey they lay up are not sufficient, the search is on for more abundant sources of sweetness. Sometimes the answer is not to find more colourful flowers: a completely different opportunity might come along in the form of a weak hive in the neighbourhood.

Scouts test the targeted hive's ability to defend itself, and if it has been weakened by parasites or pesticides, the call goes out to attack. An intense battle ensues at the hive entrance, but the defenders can't hold off the invaders for long. At some point the attacking force is just too strong, and the aggressor bees stream past the last dying fighters and advance inside. They pounce on the honeycombs and rip off their waxy covers. In no time at all they pump their honey stomachs full and fly home, carrying with them the good news for other members of their hive that there's more than enough food stored here. In the hive belonging to the weak colony, there's a loud buzzing from the incessant beating of thousands of wings as the plunderers fly in and out. When there's nothing more to collect, absolute silence descends. Unfortunately, just such a drama played out in my yard, and when I lifted the lid of the vanquished hive, what lay before me was a scene of utter devastation. Crumbs of wax on the hive floor were all that remained of the ripped and shredded combs. A few dead bees were lying around. That was it.

Despite the devastation, the attackers were far from satisfied. They had learned that life is much easier when you attack and plunder your neighbours. If the opportunity arises, they'll do the same again with another hive. As a bee-keeper, all I can do is separate the belligerent bees by setting up one of the two warring hives a few kilometres away to give the bees an opportunity to calm down. Of course this isn't an option in the wild, where the game continues until evenly matched strong hives keep each other in check.

It's not only bees that panic shortly before the onset of winter. Brown bears, for example, can't store food for hibernation, and they have to put on a layer of fat by eating instead. If there's not much food to be had in autumn, or when the bears are older and can't gather as much, things get difficult – for people as well as for the bears. The German film-maker Andreas Kieling told me the sad story of his colleague, Timothy Treadwell, who saw himself as the bears' friend and took no safety precautions around them. One day he was watching an old male grizzly in Katmai National Park in Alaska. It had clearly not yet put on enough fat to make it through the winter, probably because it was no longer quick enough to catch the salmon it needed. Experts consider such animals to be particularly dangerous. As usual, Treadwell had neither a weapon nor pepper spray with him. The old bear attacked and killed him. His girlfriend, who was right there watching, was horrified, and she began to scream. This 'predator call' (a cry of fear made by prey that triggers a predator's desire to hunt) signalled that there was more food to be had, and so the woman ended up falling victim to the hungry bear, too. The couple were later found buried near their tent. We have a pretty good idea of what happened because we have a recording. Treadwell's original intention had been to film the old bear, and his camera was left running. The cap was still over the lens, but the camera picked up the sounds of the couple's final moments.[75]

But back to animal wars. You can only talk about war, as we use the word, to describe conflicts in species that live in large social groups. In Central European latitudes, that means bee, wasp and ant colonies, which mount raids like we do. If, however, an animal attacks another individual by itself, then we talk of a fight, something you can see between many male birds or mammals.

Given that animals attack other animals, can they be considered heartless and evil? Sometimes it seems that way. I have two corner windows in my office and, when I look out of them, I can see an eighty-year-old birch tree that grows in front of the forest lodge. The old tree (birches don't live much past 100) has already experienced the ravages of time – or I should say, of woodpeckers. About 5 metres up, there is a natural nesting cavity that has been used by a variety of different birds over the years. After the woodpeckers there were nuthatches for a number of years, and then one day starlings. These lightly speckled birds began to success-fully rear their young in the cavity. One day I heard a loud screeching and, when I looked out of the window, I saw a magpie that kept flying at the tree. Suddenly it landed on the edge of the cavity and grabbed a little starling. It dropped it from the tree to the ground and began pecking away at it. Instinctively I got up, left everything in the office and rushed outside. The magpie flew a few metres and let go of its prize. The young starling was completely distraught, but it did not seem to have suffered any grave injury. I got a ladder and carefully returned it to the nest. As far as I could tell, there were no further attacks, and the baby bird got its start in life along with its siblings.

And yet things probably didn't turn out as they should have, and the fault was mine. What right did I have to get involved with the altercation? Okay, I had felt sorry for the little starling, and I couldn't just watch as it was killed. But from the point of view of the magpie, wasn't it just a piece of meat that was urgently needed to feed the magpie chicks? What if one of them had starved as a direct result of my intervention? In that moment when the magpie pulled the young starling out of the cavity, I thought it was evil. But was it really? And what does that really mean – evil? Can a quality like that depend on your point of view? If so, then from the magpie's point of view, I was the evil one who had prevented the mother or the father from making a kill. From the point of view of its species, the behaviour of the handsome black-and-white bird had been beyond reproach. But I, too, am a typical representative of my species, for most human observers finding themselves in this situation would also have taken pity on the starling.

So what would it look like if a similar incident were to take place between animals of the same species? This isn't unusual in the wild, as we can see with brown bears. In this case, it's the males that can be deadly for the cubs. When mating time approaches, males look for females who are ready to mate. Females with cubs aren't in the mood, and so males often just go ahead and take care of this. They kill the cubs, and shortly thereafter the mothers are ready for their next pregnancy – a natural response to the emergency. Because mother bears are well aware of the males' intentions, they try to keep their distance from potential admirers. Alternatively, they mate with as many males as possible, so that each thinks he might be the father of the amusing tots and leaves mother and children in peace. Scientists

from the University of Vienna have established that
promiscuous behaviour is not purely for sexual pleasure,
but is indeed a defensive strategy employed by the
females. After observing bears for twenty years in
Scandinavia, they confirmed that this behaviour happened
most often in populations where a particularly large
number of cubs were falling victim to attacks by males.[76]
     Are these male bears evil? The dictionary defines
'evil' as 'morally bad, reprehensible'. Or to put it more
clearly you could say: for an action to be evil, the perpe-
trator must intend to violate a moral code, to the detri-
ment of others. Neither magpies nor bears do that, for
their actions are part of the normal behavioural repertoire for their respective species. However, the behaviour
of the white rabbits we acquired one day was definitely
not normal.
     We wanted to switch from the typical common or
garden mixes we had on our little farm to pure-bred
animals, and we drove a few villages farther on to take
a look at white Viennas. These rabbits have soft, silky
fur and adorable blue eyes, and we just had to take a
few of them home with us. Back at our forest lodge they
got a spacious run, but the idyll lasted for only a few
weeks. One day we came into the pen and saw a wretched
little shape hunkered down on the ground. It was a
small female, and her ears were slashed so badly that
they hung in tatters. We felt so sorry for her, and we
thought that there must have been a fierce struggle for
dominance. Over the next few days there were more
and more rabbits with slashed ears and, as we watched,

we confirmed what we had begun to suspect. One of
the females had inflicted these brutal wounds on the
others with her razor-sharp front claws. Logically, the
ruthless lady was the only one still hopping around with
her ears intact.

Now, was this a bad rabbit? I believe it was. This
wasn't species-appropriate behaviour, and it couldn't be
defended on moral grounds, either. Whichever way you
look at it, the rabbit's intentions were evil. After all, the
animal must have acted intentionally, and the others had
done nothing to encourage her to act the way she did.
You could argue that this rabbit might have been traumatised by a dreadful experience in her youth that could
have driven her to behave this way. Absolutely. But isn't
this almost always the case with human evil-doers as
well? Every evil act can be traced back to a point where
it can be explained and therefore excused. For simpli-
city's sake, let's apply the same standards to both animals
and people and assume that both have at least a basic
level of free will when it comes to deciding how to act.
It's not only people who have this freedom of choice;
lots of animals do, too.

Hey, Mr Sandman

A REAL SUMMER FOR ME INCLUDES SWIFTS. They're like swallows, but much bigger and much, much faster. Uttering their piercing calls, they race between tall city buildings, hunting for insects or just for fun. Unlike many other birds, swifts spend almost their whole lives in the air. They are so extremely adapted to life above the Earth that their legs have atrophied and their tiny feet are good only for clinging onto things. Of course they also need to breed, and their nests, which they build in cliffs or in cracks in walls, are constructed so they can fly in and out of them with ease. Apart from the time they spend sitting on nests, the birds satisfy all their other requirements on the wing. Even mating often takes place high up in the air, where swifts, like us, abandon themselves to the moment. And having a male clinging to the female's back does nothing for the birds' flying prowess, so mating pairs often end up spiralling down at breakneck speed and have to separate in time to avoid being smashed to smithereens.

But I wanted to introduce you to the swift to illustrate another bodily function: sleep. Most life forms (even trees) need to sleep, and birds land in a sheltered place to do so. Our chickens, for example, take themselves back to the hen house at dusk like good little chickens, tramp up the ramp and settle on their perch, where they snuggle up side by side. They don't have to worry about falling off, because, as with most birds, their tendons shorten when they sit and their toes curl automatically. This means our chickens can hold on tight without expending any energy. Like all birds, chickens dream. When they do, there's a danger they might move around, just as we do, as their nightly movies play. But if they were to move, they might fall from their perch or, for birds living in the wild, out of their tree. And that's why the muscles used for movement are deactivated the moment a bird nods off, so it can spend the night peacefully with its head tucked underneath its wing.

And what about swifts? They never perch, and they spend not a second longer than they have to on the ground or on the nest. If they want to sleep, they do so while airborne. That is highly risky, of course, because sleeping birds aren't in total control of their actions. And so they spiral upwards a couple of kilometres to increase the distance between themselves and the ground. Then they begin to glide downwards, tracing a wide circle that slows their descent. Finally, they are free to doze for a few moments. They don't have time for anything more, because they need to be wide awake again before the first rooftops loom dangerously close and their situation becomes precarious.

Is this brief shut-eye sufficient for the birds to get any rest? Definitely, because although sleep allows all species to exclude or reduce outside influences so that the brain can run its internal processes undisturbed, sleep is a little different for every species. The different phases of our sleep, with their varied depths, show that even human sleep is not a uniform affair. And our horses, for example, don't need much in the way of really deep sleep. Often just a few minutes are enough, which they take while lying down on their sides looking as though they've been shot. They're so deep in dream-land that they are indeed dead to the world, and their legs twitch as though they were galloping over an imaginary prairie. Other than that, they stay on their feet and doze away a few hours of each day just like the airborne swifts.

It's obvious that, like us, animals sleep. Even tiny fruit flies need to sleep, and when they're asleep, they twitch their legs just like horses. The really exciting question is how they sleep and what they dream about. Our nightly mental excursions take place during the so-called REM phase of sleep. REM stands for 'rapid eye movement'. In this phase, our eyes move under our eyelids, and if you wake a person in REM sleep, they can almost always remember what they were dreaming about. Many species of animals experience similar nightly eye movements, and the larger their brain in relationship to their body size, the more they have. Because animals can't talk to us, we have to look for other clues to understand what's going on inside their heads. Researchers at the Massachusetts Institute of Technology in Boston looked at rats. They measured the rats' brainwaves while they were in a maze busily searching for food. Then they compared these readings with instrument displays while the rodents were sleeping. The similarities were so clear that, using the data they had gathered, the researchers could even tell where in the labyrinth the sleeping rats were in their dream.[77]

Experiments with cats in 1967 also yielded evidence for dreaming, although in this case the evidence was indirect. The scientist Michel Jouvet from the University of Lyon stopped cats from relaxing their muscles while they slept. Normally the body, including the human body, shuts off voluntary muscle movement to prevent us from thrashing about wildly in our dreams or walking around our bedrooms with our eyes closed. This shut-off mechanism is only necessary in dream sleep. When the mechanism is deactivated, you can observe what the test subject is experiencing in its dreams. Jouvet observed cats in this state arching their backs, hissing or running around, all while in a deep sleep. Science accepts this as proof that cats dream.[78]

But what does it look like if we distance ourselves from our branch of the animal kingdom, leave the mammals and consider insects instead? Could something similar play out inside such tiny heads? Can the relatively small number of cells in a fly's brain also produce pictures during sleep? Today there are indeed indicators that these tiny clumps of cells can do more than we ever gave them credit for. As I just mentioned, fruit flies twitch their legs just before they fall asleep, and their brain is particularly active while they are sleeping – another parallel with mammals. Does this mean that fruit flies dream? Their physical reactions suggest that they do, but so far we can only guess what images might be lighting up inside their little heads. Pictures of mushy fruit, perhaps.[79]

Animal Oracles

I HAVE TO ADMIT I always used to be a bit sceptical when there was talk of animals having a sixth sense. Okay, so in many species individual senses are more pronounced than ours – but can they really be so much more pronounced that they can pick up almost imperceptible clues to impending natural catastrophes? Today, I believe that this sixth sense is a necessary tool for survival in the wild, a tool that has not been completely lost in the artificial surroundings of our modern world, but which certainly has been buried.

The word 'buried' is the clue here, for who would want to be buried alive in a volcanic eruption, for example? Goats, it seems, are particularly afraid of this eventuality, or so you might think, given their ability to detect volcanic activity. A researcher at the Max Planck Institute, Martin Wikelski, discovered this by tracking the goats. After he fitted a herd of goats on the Sicilian volcano Mount Etna with GPS transmitters, he noticed there were several times when the herd suddenly became restless as though a dog were harassing them. The animals ran back and forth and tried to hide under bushes or behind trees. Each time, a few hours later there was a sizeable volcanic eruption. The herd gave no noticeable early-warning signals for minor eruptions – and, indeed, why should they bother? How were the goats aware of what was about to happen? Unfortunately, researchers don't yet have a satisfactory answer to that question, although they believe it has something to do with the gases seeping up out of the ground just before an eruption.[80]

Animals in German forests are aware of similar dangers. Volcanic activity is an issue everywhere in Central Europe, and you can see signs of it in the Eifel, where I live. Ancient volcanoes tower over the land-scape, interspersed with a few young ones, like the one that formed Laach Lake. Young, in this context, means that the last time the volcano erupted was 13,000 years ago, and it could do so again at any time. Back then, it spewed 16 cubic kilometres of debris and ash up into the air, shook Stone Age settlements and formed dust clouds that obscured the sun as far away as Sweden. So it's a danger to be taken seriously, even if there's only a slim chance that we'll experience an eruption in our lifetime.

Where I live, wood ants are the focus for researchers – or I should say for some researchers. Professor Ulrich Schreiber from the University of Duisburg-Essen and his team went to incredible effort to map more than 3,000 anthills in the Eifel. They discovered a striking correlation between the placement of the anthills and faults in the Earth's crust caused by volcanic eruptions and earthquakes. At the intersections of these lines of disturbance, they found concentrations of anthills. The composition of the gas seeping out of the ground in

these places was distinctly different from what consti-
tuted the surrounding air. The red wood ants love the
mix and prefer to build their homes at these junctions.[81]
Now that I'm aware of this, I think of it every time I'm
out in the woods and see these attractive formations
full of scurrying ants. Scientists don't have the answer
to the goat riddle yet, and neither do they know why
the ants like these places. The one thing that is clear
is that the ants can smell the tiny differences in the
concentrations of gases, just as the goats can, and there
are numerous reports from all over the world about
similar phenomena.

Does this mean animals are inherently more sensitive to their surroundings than people? Naturally there
are many species that perform considerably better in
individual sense categories. Eagles see better, and dogs
hear and smell better than we do. However, the sum
total of our senses is so good that on average we are
no different from other species. So why is it that, in
comparison with animals, we are so unaware of changes
in our environment? I believe the answer lies in the way
our modern home and work environments overwhelm
our senses. Most smells, for example, no longer come
from woodlands or meadows, but from exhaust pipes,
emissions from the office printer or the perfumes and
deodorants we use on our bodies. Permanent sensory
overload, as a result of fabricated fragrances, masks
natural aromas. It's different out in the country only if
you spend a lot of time out and about in Nature. For
instance, where I live you can smell a single moped

puffing and discharging exhaust from its stinky two-cylinder engine from 50 metres away and, when it rains, the forest air is immediately full of mushroomy smells that forecast a rich harvest in a few days.

It's similar when it comes to keen vision. People who spend a lot of time sitting in front of a computer screen when they're young, or surfing with smartphones, are more likely to be short-sighted than children who spend most of their time outside. Recent studies show that short-sightedness has increased significantly in younger generations, with rates up to four times higher in people born in the 1960s compared with people born in the 1920s.[82] Are we losing our distance vision? Luckily we can put on glasses; however, the increasing deterioration of our natural acuity of vision seems to me symptomatic of something else.

I believe the senses we are born with are as attuned to natural processes as animal senses are. It is modern life that dulls one sense after another. Even my ears aren't the best these days, as a few frequencies have fallen victim to disco visits in my youth or to shooting practice. But there is still hope. When something is gone organically, it cannot be repaired; however, our brain can compensate for many things. For me, one beautiful example is the annual migration of cranes. I can hear these birds from a long way off, even through windows well insulated for sound, because I'm always excited to hear these ambassadors of seasonal change. All it takes is a faint suggestion, more like a premonition, and I step outside and see a V of birds flying in the distance.

Migrating cranes are closely tied to the subject of this chapter – the early-warning systems of animals. The birds are clues to distant weather patterns, because they like to fly with the wind at their backs. When they fly down from the north in autumn, it means bitterly cold north winds are blowing south, which might signal the first snows of the season. In spring, however, the mass arrival of cranes signals the start of the breeding season, meaning that warm southern breezes are drifting north from their winter quarters in Spain and temperatures will soon be on the rise.

You can even roughly estimate the current temperature by listening. That sounds bizarre, but it's actually quite simple. Grasshoppers and crickets are our little helpers here. These cold-blooded insects begin their concerts when the temperature reaches about 12 degrees Celsius. The higher the thermometer climbs, the faster they chirp. You could argue that a better way to tell the temperature would be by how hot the air feels on your skin. True, but when you're physically active that's difficult, because of the additional heat generated by your own body.

Just as you can train your ears, so you can also train your eyes. You can correct defective eyesight with glasses, but it is the brain's reaction that is even more important, and you can hone its ability to pick up on particular changes in the surroundings. These days my peripheral vision picks up deer simply by sensing an aberration in the green of the trees. The slightest difference in colour in spruce attacked by bark beetles catches my attention before clear differences appear between affected trees and the healthy crowns of neighbouring ones.

Other senses of mine have been trained, as well. Whether I can feel a shift in the direction of the wind blowing on my face, signalling a change in the weather, or the small droplets of rain that announce light cloud cover (and therefore no serious rain), or minuscule anomalies in the smell of the air that betray the presence of a rotting animal carcass in the distance – together these sensations are pieces of a puzzle that constantly update me on the current state of my surroundings and its dangers, without me having to think too much about it.

If you belong to that segment of the population that has a feel for the weather, you too could forecast a storm long before the first clouds appear in a bright-blue sky. Scientists don't agree where this sensibility lies – for example, whether it has something to do with the changing conductivity of cell membranes – but whatever it is, it works. How much more accurately must early peoples have been able to read the woods and the meadows, exposed, as they were, to all those stimuli day in and day out? In my case, I spend only part of the day training my senses like this. Animals train all their lives. No wonder they are so much better at predicting natural disasters.

If animals are so sensitive, what about long-term forecasts? Can animals feel if the coming winter will be a hard one? Some years, you can see squirrels and jays burying particularly large quantities of acorns and beechnuts. I'm sorry to have to tell you that it would be wrong to conclude they are acting with wise foresight to tide themselves over a long and snowy winter. The animals are simply making the most of an overly abundant offering of food that the trees have put at their disposal. Beeches and oaks all bloom together every three to five years. This synchronous blooming often happens the spring after a very dry and difficult summer. The bountiful nuts and the squirrels' and jays' diligent nut-gathering happen later that year, which means that their activity attests to what the weather was, rather than what it will be, which is a bit disappointing.

So animals are not capable of forecasting the weather in the long term. However, the situation is quite different when we turn our attention to short-term changes. And my favourite species in this respect is the chaffinch. These birds like to live in old deciduous forests, but they also hang out in mixed woodlands. There the male trills a pretty song with a rhythm that I remember learning in school, which goes something like 'chip chip chip chooee chooee cheeoo'. But you can only hear this song in good weather. If dark clouds are massing – the first drops of rain might even be falling – then what you hear is 'run run run run run', sounded on a single note. I've noticed during my daily forest rounds that the chaffinch changes its song when it's disturbed, but it keeps singing its regular tune when it sees me. It clearly registers the sudden appearance of a person from behind the trees as less of a threat than the sudden disappearance of the sun behind storm clouds.

But what benefit do other chaffinches get, when one alert bird notices a change in the weather and warns the others? Couldn't they just all look up into the sky and see the storm front for themselves? Not if they are under the dense leafy canopy of an old beech forest. Here the most they would notice is a slight darkening of the day. The only place to see the approaching menace is through a gap created when one of the giant trees fell and opened up a view of the sky, or from way up in the crowns of the trees. Thus the warnings are useful because not every finch is privy to this 'insight' from where it is sitting.

Animals Age, Too

IT'S WIDELY KNOWN THAT ANIMALS get a pain here and an ache there with increasing age, just as we do. But what's going on inside their heads as they become increasingly frail? Are they aware of their diminishing physical capabilities? This question is hard to answer directly using science, but, once again, we can get a good idea by observing their behaviour.

Horses seem to get more fearful as the years advance, and with good reason. As I've already mentioned, normally horses do very well dozing standing up. They even have a special knee joint to help them do this. The joint locks in place when the muscles relax, to prevent the leg from buckling when the horse is asleep. When the horse dozes, it alternates hind legs, switching its weight to the locked leg, while it rests the tip of its other hoof on the ground. This takes the weight off the forelegs, which both remain straight. A horse can doze for hours at a time like this, but it's not real sleep. Just like us, horses need true deep sleep to stay healthy and fit. And to have that kind of sleep, they must lie down on their side with their legs stretched out on the ground. Only then can they drift off into the land of dreams, with heightened brain activity and twitching hooves. Sometimes their lower lip also moves, as though they were whinnying in their sleep or trying to eat. Once their slumber is over, they have to get back up again. As they have relatively long legs and may weigh more than 500 kilos, this manoeuvre requires a great deal of strength. First they heave themselves up with their front legs extended; then they push off the ground with their hind legs and stand up.

Creating the momentum to get up off the ground is almost impossible for old horses, and you can see how the animals get really scared of lying down. Even though they would love to relax completely lying on their side, for safety's sake they remain standing and content themselves with dozing. That's definitely not good, because energy reserves are drained more quickly in the absence of any real sleep. But the animals clearly know very well that by lying down they are putting themselves in a life-threatening situation. A horse that cannot get up again will soon be dead, because its inner organs will cease to function properly – or a predator will come along. As getting up gradually becomes more difficult, the amount of deep sleep the horse gets diminishes at the same slow rate. We can see this with our two mares. The older one, who is now twenty-three, can be found lying down considerably less often than her companion, who is younger by three years. One day it will get to the point when fear wins – and then Zipy will never dream again.

You can also see changes with old female deer. Apart from the loss of muscle, which makes the does look bonier, there are changes in their behaviour. Let me just come right out and say it: they get grumpy and argumentative. That's understandable, given that they might once have led the herd and been revered as royalty. They are still able to bear young in their old age, but the fawns they bring into the world are weak. The old does' teeth have been worn down completely by years of use, so they can no longer chew their food properly and therefore they are often hungry. This means they are scrawny and their milk production drops, as does its fat content, which is produced in the udder, and so the old does' offspring are starving, too. It's no wonder these fawns are the ones that most often fall victim to diseases or carnivores. And this, in turn, affects the status of the ageing deer, as I explained earlier. Under similar circumstances you'd probably be in a bad mood, too, wouldn't you?

One subject I've hardly ever read anything about in ageing animals is dementia. Our house-pets certainly live much longer than they used to, because their medical care has kept up with ours. Our little Münsterländer is a good example. She always had the best in food, she got her injections, she was taken to the vet for treatment when she had an infection and, while she was there, she had tartar removed so that she didn't lose her teeth. One day, at the age of twelve, Maxi began to stagger. The problem was quickly diagnosed: she'd had a stroke. That hit us hard. Our little dog, who'd always been so spry, was suddenly nearing the end of her life. But the prescribed pills and injections worked quickly, and Maxi recovered.

Maxi's faculties and senses gradually diminished as she lived out an appropriately long phase of old age. At some point she simply stopped making any noise – barking became a thing of the past, which didn't bother us. Quite the opposite, actually. To match that, her hearing then left her, which was a bit more burdensome, because it meant we could only communicate with her when she could see us. But our dog was still a creature that enjoyed life. In her last year, however, she began to lose it, until she no longer knew who we were. She also spent hours turning around in her basket as though she was going to lie down, but she never did. She then ate less and lost a great deal of weight, and cancerous growths appeared. It was with a heavy heart that we asked the vet to put her down. Her successor, the cocker spaniel Barry, went through a similar process at the age of fifteen. Apart from the loss of his mental faculties, he also became incontinent, which cost us a great deal of work and enormous quantities of carpet-cleaning foam. You can now find therapies and medications to counteract what is technically referred to as 'cognitive dysfunction'.

I believe all higher animals can suffer from dementia. Cat lovers report similar behaviour in their beloved charges, and researchers have found deposits and changes in the brains of these house-pets that are similar to those discovered in people suffering from dementia. We even had a goat in our herd with dementia. It lost its ability to orient itself and one day, thanks only to our son's diligent efforts, we found it lying peacefully in a stream in the woods. Observations in the wild are few and far between, because animals suffering from dementia are easy prey for carnivores. They isolate themselves from the community of the herd and, in doing so, signal their vulnerability. An animal that is no longer right in the head is mercilessly eliminated. Of course, the same thing happens with predators, only instead of falling prey to other animals, they starve.

But what does it look like when the end is approaching and the grey cells are still fully functioning? Are these animals aware that the end is near? Not many people, but at least some, can foresee their death. Some are ill and predict the time of their demise almost to the week; others are old and tired and simply don't want to live any more. For these people, death comes as no surprise. It's the same with some animals. Our very old female goats separate themselves from the herd shortly before their time comes, so that they can die in peace. When they go into seclusion, they must know what's in store for them. They search out an isolated patch of pasture or the small shelter the herd doesn't use during daytime in summer, and here they lie down and die peacefully.

How do I know this? You can see it from the position of the deceased animal. Our favourite goat, Schwänli, for example, settled down comfortably on her stomach with her legs tucked under her. This is a normal relaxed position for goats when they sleep. In contrast, when an animal dies in distress, it thrashes its legs, disturbing the earth around it. It's on its side, its neck is bent back and its tongue is often hanging out. From the animal's position, you can tell that it suffered in its last moments. Not our Schwänli, though. She had clearly anticipated her death and quietly took leave of this life. Such behaviour doesn't only make saying goodbye easier for us; it also has advantages for the herd, at least in the case of wild animals. Weak old animals present a danger, because their slow movements attract predators. By separating themselves from the herd in a timely manner, they make sure that younger members of the herd won't be torn apart as collateral damage when the predator comes for them.

Alien Worlds

WE OFTEN FIND NATURE IDYLLIC and calming because it looks so peaceful and harmonious. Colourful butterflies flit through flowery meadows, the white trunks of birches tower over the undergrowth as their leaves wave in the breeze. It really is pure relaxation for us, partly because today's wild holds few dangers for people. That is not the case for the creatures that live there, however, and so they view this idyll from a completely different perspective.

If you look, you'll find two distinct differences between butterflies and moths. Butterflies are beautifully coloured. The peacock butterfly, for example, sports a large eye on each wing to scare off birds and other hungry creatures. Its body and wings are relatively hairless, so the image the attacker sees is shiny and clear. Moths, in contrast, tend to the monochrome. Grey and brown are their preferred colours, because they spend their days dozing on bark and branches, waiting for dark. During the day they're sluggish and could easily fall prey to birds, which can distinguish any variation in colour with their sharp eyes. Woe to the moth whose wing colour doesn't match the bark it's resting on, because that means it's landed on the wrong tree. Moths lacking in tree-identification skills don't survive the day – or, from the moth's point of view, the night.

In order to survive, moths and butterflies even adapt to a world transformed by human activity. Let's take the peppered moth, for instance. It has a wingspan of 5 centimetres and black specks on its white wings. That's exactly the colour of the birch bark on which it likes to rest. However, in England birch trees were white only until about 1845. After that, the Industrial Revolution and the burning of coal released so much soot into the air that a grimy black layer built up on the bark. The moths that used to be so well camouflaged now stood out, and hundreds of thousands of them were eaten by birds, with the exception of a few misfits. These creatures had always been around and, like black sheep, their wings were dark – a death-sentence until then. But now the dark moths turned out to be the winners. They were the ones that survived and were the reason that, after a few years, most peppered moths were black. It was not until the end of the 1960s, when measures were taken to improve air quality, that the situation changed. The birches became cleaner, which meant they were now white again. And that's why by 1970 most of the peppered moths that were observed were white again.[83]

At night, things literally look different. Colours are mostly unimportant because insect-eating birds sleep the night away perched on the branches of trees. Other hunters arrive on the scene: bats. They rely less on sight when they hunt and more on ultrasound. They make high-pitched calls and then listen to the echo sent back by objects and potential prey. Visual camouflage doesn't help one little bit, because the flying mammals are 'seeing' with their ears. Therefore the moths must make themselves invisible to hearing. But how do you do that? One possibility is to absorb sound, instead of reflecting it. And that's why many moths are covered with a thick furry layer that traps the bats' calls or, to put it more precisely, muddles it up and reflects it back all over the place. Instead of receiving a sharp image of a moth, the bat's brain gets a fuzzy something that might just as easily be a bit of bark.

Pigeons also see completely differently from the way we do. It's true that they are visual animals like us, relying heavily on sight, and they need daylight to see. However, in addition to all the details that are part of our lives, they are clearly aware of other things in the air. They see the polarised pattern of light in the sky – that is to say, the geometrical orientation of oscillating light waves – and this polarisation is directed north. This means that by day pigeons see a compass everywhere they look. No wonder carrier pigeons orient themselves well over long distances and always find their way home.[84]

Once we include bats' hearing as 'seeing,' we can widen the range with other species as well, to understand what they feel and the subjective world they live in. With dogs, for instance, we can wonder how strongly their sense of sight, which is somewhat less acute than ours, is enhanced by their senses of smell and hearing.

If it is the sum of the impressions that completes the picture, then we don't know what dogs see, if we pass judgement on their eyesight alone. If we did that, we'd have to say that dogs are in dire need of a pair of glasses, because the lenses in their eyes don't do a good job of adapting to different distances, meaning that dogs see objects in focus only when they approach within a range of about 6 metres. If they get closer than about 50 centimetres, they lose focus again. Dogs use about 160,000 visual nerve fibres to create these images, whereas in our eyes there are about 1.2 million.[85] But even with 'visual animals', sight on its own is not enough, and you can test this for yourself right now. If you happen to be sitting somewhere noisy, with conversation or street noise in the background, just put your hands over your ears for a moment. That you can hear almost nothing is not the point. The point is that the three-dimensional feeling of your surroundings abruptly changes. Depth is lost. To what extent might the picture that dogs have of the world depend on their ears, which are fifteen times more sensitive than ours?

I find it endlessly fascinating when I think that every species of animal sees and feels the world in a completely different way, so you could say there are hundreds of thousands of different worlds out there. And many of these worlds are waiting to be discovered, even in the latitudes where I live. Apart from the species I've introduced already, there are many thousands of others in Central Europe that are unfortunately so minute and unattractive that even their distribution has not yet been systematically researched. Regrettably, nothing is known about their feelings, either, because if they don't have a recognised relevance for us, little money is forthcoming for research. And when we don't know what's going on inside these little animals, what their needs are or how they suffer under commercial forest practices, no one is interested in designating conservation areas for them.

I, for one, would be really interested to know more about weevils, for example. This family includes flightless species that have captured my heart. One representative is a tiny brown nipper, only 2 millimetres long, that looks just like a tiny elephant. Its hair grows in stripes along its head and back, as though it were a devotee of the Mohican hairstyle. Weevils are adapted for life in the rotting leaves of ancient forests, and these forests are distinguished by one thing above all: they hardly ever change. In Germany these forests are mostly beech. These trees create a stable community and actively support one another. They exchange sugar and information so efficiently through their interconnected root systems that storms, insects and even climate change hardly affect them at all. It's an easy place for beetles to live as they munch their way peaceably through the dead leaves on the forest floor. These beetles are a relic species dating back to the primeval forest. They have survived for uncountable generations and, if you find them, you can be sure the deciduous forest is centuries old. Why would they wish to travel anywhere else? What use would they have for wings? They don't

need to find a new place to live. Thousands of generations can grow old here undisturbed, and I'm happy to
report that the forest I manage is one of the places
where one species of these fascinating beetles has been
found. I have to say that they grow old by beetle standards, because it takes less than a year for these little
animals to enter their dotage.

Without wings, weevils cannot fly away and,
between birds and spiders, they have more than enough
predators. If you're afraid but you can't run or hide, you
have to come up with other strategies, so when they're
disturbed, weevils simply play dead. Thanks to their
patterned leaf-brown camouflage, they're very difficult
to spot. Unfortunately, they're also difficult for visitors
to the wood to find unless they have a magnifying glass:
weevils are measured in minuscule fractions of centimetres. In the absence of ongoing research, we can only
speculate what feelings these little chaps might have,
other than fear. Despite the lack of data, it was important
for me to mention them as an example of the many
species that are not the focus of our attention and yet
still deserve recognition. For the diversity of life that
surrounds us truly is amazing: colourful birds, cuddly
mammals, fascinating amphibians and even useful earthworms. There are interesting animals to be seen everywhere. And that is exactly our Achilles heel. We only
admire what we can see, but the majority of creatures
in the animal world are so tiny that they are revealed
to us only with the help of a magnifying glass; and,
especially for the tiny ones, we need a microscope.

What about tardigrades (also known as water bears), for example, of which more than 1,000 species have been discovered so far? With eight legs and a cuddly body – they really do look like little bears with too many legs – these eumetazoans (belonging to the scientific sub-kingdom Eumetazoa) like it to be really moist. The species native to Central Europe mostly range between 0.25 and 0.5 millimetres in length and they prefer to live in moss, which also loves water and stores it particularly well. The tiny 'bears' scuttle around here and, depending on the species, eat plant material or hunt even smaller life forms, such as roundworms. But what happens when their homes dry out in the hot summer months?

In the forest I manage, the lush green moss at the bottom of thick beeches is often brown and crispy dry, come summer, so the little bears have absolutely no access to water. They then fall into an extreme form of sleep during which they dry out. Only well-nourished tardigrades survive, for fat plays an important role. If moisture is lost too quickly, death follows; however, if moisture evaporates gradually, the tardigrades adjust, dry out, draw their tiny legs up into their bodies and reduce their metabolic rate to zero. In this state of suspended animation they can withstand almost anything: neither searing heat nor bone-chilling cold can touch them. Absolutely no biological activity takes place. They do not dream, because that inner projector requires energy to roll. You could say it's a kind of death, which means there's no ageing, either. In the general scheme of things, tardigrades are not long-lived, but under extreme conditions they can survive for decades, waiting for rain to reanimate them. When rain comes and saturates both the desiccated moss and the tardigrades, it takes no more than twenty minutes for the tiny creatures to extend their legs and get their internal structures back on track. Life, as they know it, resumes.[86]

Artificial Environments

EVERY DAY WE TRANSFORM our planet in some way, and every day we lose more of Nature in its original form. We have already cleared, built on or dug up an unbelievable 80 per cent of the Earth's land mass.[87] Animal senses, however, are not configured for concrete and tarmac, but for woods, moors and intact waterscapes. Artificial light is just one example of the many ways in which we confuse them.

In Europe at least half the night sky is affected by light pollution. Even a small town with 30,000 inhabitants contributes to an artificial brightening for 25 kilometres in every direction. The people who live there have little opportunity to observe a pristine starry sky. And it's not just people that are affected. Many species of animals, especially insects, depend on stars to orient themselves when they travel at night. Moths, for instance, rely on the moon when they want to fly in a straight line. For example, when the moon is at its height and they want to fly west, all moths have to do is keep the moon to their left. But little moths can't tell the difference between the moon and a cosy lamp adding a decorative touch to a garden at night. Now, as the tiny winged wanderer glides past the tulips and the roses, it gets disoriented. The brightest light at night must be the moon, mustn't it? And so it tries to keep this new moon to its left, but the lamp is unfortunately not 384,400 kilometres away, but only a few metres. If the moth keeps flying in a straight line, the 'moon' appears behind it, and it seems to the moth that it must have flown in a circle. And so the insect pilot corrects its course to the left in order to, as it thinks, continue flying straight ahead. This makes the 'moon' appear on the correct side, but what's really happening is that the moth is flying in circles around the light. The spiralling flight takes the moth ever closer to the light until it finally ends up at the centre. If the artificial moon is a candle, there's a brief 'puff' and the moth's life is snuffed out.

But even without this dramatic finale, the moth is doomed. If it tries to fly a straight course all night and keeps ending up at the light, at some point it will have exhausted its energy reserves. Its intention was to fly to night-blooming plants to fill up on nectar, but the few hours it has left to feed have morphed into an involuntary weight-loss programme. And if that wasn't nightmare enough, predators have adapted their behaviour to take advantage of the new normal. Spiders spin webs under the lights by our front door, because this is where they hit the jackpot. The moth starts its spiral flight around the light and continues until it inevitably lands on the sticky strands, where it is dispatched by the toxic teeth of the owner.

Roads are particularly challenging for wild animals. First off, tarmac is not in and of itself something

negative, because insects and reptiles can warm themselves here until their body temperatures are high
enough for them to function properly. Dark surfaces
heat up particularly well, which means that, especially
in the cooler days of spring, cold-blooded animals (which
can generate very little heat themselves) are able to get
a jump on the day. That is, as long as a passing car
doesn't brutally end their sun-worshipping. Roads are
attractive in other ways, too. Take deer, for example.
Many roadside verges are mown regularly, which means
they are a good source of tender grasses and other
toothsome greenery. And they are particularly safe
places for game animals to hang out, as hunting is
forbidden there so as not to endanger drivers. No wonder
you can see an impressive number of game animals in
these unusual habitats at night. Unfortunately that,
along with the fact that there are so many game animals,
leads to a large number of traffic accidents. The German
insurance industry reports about 250,000 collisions
involving wild boar, deer and other wild animals a year
– often with deadly consequences for the animals.[88]

Animals should be capable of learning. They should.
But there are two main reasons for the constant supply
of crash victims. The first reason is youthful careless-
ness, which is not restricted to humans. For example,
when deer pass their first birthday, they set off to find
territory of their own. While old-established deer often
move no more than 100 metres in a whole day, which
they spend snacking on tender raspberry leaves, young
animals are chased away until they chance upon

unoccupied space. And, given a density of 646 of regional
roads per square kilometre of countryside, the yearlings
have to cross many bands of tarmac before they find a
peaceful unoccupied corner they can call home.

The second reason is love. Roe-deer bucks in
particular go completely berserk at mating time and
have only one thing on their mind: sex. In the heat of
the summer months of July and August hormones run
amok, and the male deer are constantly straining to
hear a seductive peep. Female deer ready to mate make
this sound to draw attention to themselves. Because
hunters imitate this call using a blade of grass or a leaf
(you hold the blade or leaf tightly between your thumbs,
bring your mouth close and blow), the rutting season
is also called 'leaf time' in Germany. I admit that even
I have tricked a roe buck this way, because I wanted to
see if it really worked. And it did: right after the first
soft peep, a yearling sprang out of the undergrowth
looking around to see where the lady of his dreams
might be standing. Because the bucks' senses are in a
state of complete confusion, when an amorous adventure
beckons, they jump out onto the road without even
bothering to look. And that's why in the summer more
of the collisions with wild game in Germany involve
roe bucks, by day as well as by night.

Does this mean that cities are bad places for wildlife? Not at all. Apart from the restrictions and dangers
I've mentioned, there are also great opportunities, above
all for species diversity. Outside the city limits, fields
and pastures are drowning in seas of liquid manure and

being transformed into wastelands, and in the forests clear-cut logging machines are sawing down one tree after another and compacting the soil beyond repair, while in the cities, between the rows of houses, new, relatively intact habitats are appearing. No wonder a large number of species from the decimated agrarian deserts have fled to these refugia, including thousands of plants. Scientists estimate that about 50 per cent of the native regional and national species of plants in the Northern Hemisphere are to be found in cities. This means our suburbs are now becoming hotspots of diversity. Why am I highlighting the distribution of plants, in a book about animals? Well, herbaceous plants, shrubs and trees are food for animals. They are the first link in the food chain, and therefore they are important indicators for habitat quality. And this means that there are encouraging findings to report for animals as well. For example, 65 per cent of all the species of birds in Poland can be found in Warsaw.

Cities are budding natural areas, comparable to volcanic islands that rise out of the sea in turbulent times, naked and desolate, only to be settled by plants and animals as the years pass. What such young habitats have in common is that they are subject to dramatic changes, so in cities it may take many more decades or even centuries before competing species settle into equilibrium. In Berlin, Munich and Hamburg (the cities I am most familiar with) you can witness a steady, if slow, transformation. At first, a disproportionate number of non-native species gain a foothold, because they are 'set out' in – that is to say, introduced into – gardens and parks by the city's inhabitants. It takes many centuries for native varieties to proliferate and reclaim neighbour-hoods. You can look to the US and Italy to follow their progress. The number of non-native plants in the States decreases from east to west, mirroring the waves of settlement by Europeans; and in Rome their numbers have decreased to just 12.4 per cent of the whole. The Eternal City has had more than 2,000 years to achieve this balance.[89]

You can see a similar trajectory with animals. Generalists like the fox, which adapt to a wide range of environments, have done relatively well, but overall animals appear to have more problems settling in than plants, because they need larger territories and they are also threatened by cats, other pets and traffic. And if a species is particularly successful – pigeons, for example – then we are no longer kindly disposed towards its representatives; in some places, we even begin to mount campaigns against them. I find urban bee-keeping to be a particularly positive development. In contrast to the open countryside, there is a good selection of plants in bloom within cities all summer long, which means that the number of hives and the amount of honey they produce is steadily increasing. This shows that there must be enough food for butterflies and bumblebees as well. And so we can conclude that urban areas do not necessarily exclude animals. That being said, we mustn't lose sight of the importance of protecting their original habitats, which is another issue altogether.

In the Service of Humanity

MOST ANIMALS USED BY PEOPLE lead dismal lives. Countless pigs and chickens in factory farms are regarded as nothing more than generators of raw material. There's no need to discuss whether these animals voluntarily and happily work for us; we can surely agree that they don't. However, there are touching examples of human–animal partnerships that are a joy to behold. I often see these partnerships in action in the forest I manage. I'm talking about the log-hauler and horse teams that remove felled trees. These days, most trees are felled using a mechanical harvester. These hugely heavy machines are detrimental to the forest, because they crush the delicate soil as far as 2 metres below the surface. Therefore in my own community forest, we use forest workers to cut down trees. Felled trees then need to be dragged out along logging roads. And in Hümmel removal is done using heavy horses, just as it was hundreds of years ago.

Do these horses enjoy their work? Don't they find it tedious to spend the whole day dragging heavy loads until the sweat runs down their flanks? First, let's talk about the load. To reduce the weight, forest workers cut the trunks, which can be up to 30 metres long, into 5-metre sections. These are not only lighter, but are also easier to manoeuvre around standing trees. Then there are the log-haulers. I've never met one who didn't love his or her animals. Their horses are their colleagues, and they don't want to overwork them. Because there are no holidays or weekends when you're looking after horses, the animals are more like valued family members. When they're out in the woods, the log-haulers are careful that nothing happens to their charges, and it is the horses themselves that are eager to work just that little bit harder. You can see how much they enjoy their work when they are forced to take a break.

Log-haulers usually have a second horse ready to take over, so that they can fill a reasonable daily quota of logs. In the first half of the workday, at least, the 'resting' horse paws impatiently at the ground. It would much rather be back at work than taking a breather. Even when the horses are working, it would be easy for them to baulk, because the log-haulers usually guide them with a slack rein. The leather lines would be far too weak to stop these giants, which weigh in at a few tonnes, or to drag them in a particular direction. The reins serve mainly to keep the log-haulers in contact with their horses and to pass small signals forward. The rest is accomplished by means of an incomprehensible gibberish, babbling sounds like 'Jojo, hejhe, brr'. The horses listen and know exactly whether they are supposed to go forward, backwards or sideways, and whether they should pull with all their might or advance carefully. You can find similar human–animal partnerships with shepherds and their dogs, which also follow verbal commands. This is another example of animals taking pleasure in their work, as you can clearly see if you watch sheepdogs racing around a flock of sheep to round them up.

On the subject of 'domestic animals', there are two completely different points of view. One is that we have moulded our fellow creatures through breeding to make them perfectly suited to meet our needs. Wild has become tame, slim has become fat, large has become small – no matter what our desires, animals can meet them. Some species have been formed into bizarre caricatures of their former selves. However, you could also look at this process from another perspective: that of the animals. They have succeeded in changing themselves so that they know exactly how to push our emotional buttons. And this brings us back to Crusty, the French bulldog. This little snub-nosed dog exuded a natural charm – you just had to stroke him. So who was manipulating whom? Food and water were provided. If there was a slight injury, there was a trip to the vet. In winter, there was always a comfortable spot by the fire. The little chap led a really agreeable life. If he had still been roaming around outside like his ancestors, the wolves, there's no way his life would have looked like this.

The example of lactose tolerance also shows how far we have adapted to life with our four-legged companions. Normally only infants tolerate milk, because they are the only reason mothers produce the white liquid. The ability to digest milk, or more specifically lactose, is gradually lost as infants adapt to solid food. Or I should say it used to be lost. When people began to keep domestic animals, it became possible for adults to consume milk and cheese from their cows and goats. Because milk is a valuable food, communities where a genetic mutation made it possible for their members to digest lactose had a better chance of survival. This process can be traced back about 8,000 years and still continues, meaning that in Central Europe 90 per cent, but in Asia only 10 per cent, of the population are lactose-tolerant. There's no research yet to see how people might have adapted to live with dogs. Depending upon which scientist you ask, this relationship has been going on for as long as 40,000 years.[90]

Communication

As I HAVE ALREADY MENTIONED, in the end we will never know if fear, grief, joy and happiness feel the same to animals as they do to us. We cannot even say for certain if one person feels the same as another, as you have perhaps discovered for yourself when it comes to pain. You can use stinging nettles to test whether some people are more sensitive than others to the same experience: some scream out loud, while others barely feel anything. And yet we can talk to one another, so we can get at least some idea of what other people are feeling. We can't do that with animals.

Really? Studies of ravens tell a very different story, as we have already seen when it comes to naming. Greeting new arrivals using variously pitched calls – some higher, some lower – gives immediate feedback on the esteem in which they are held. There's no better way to express emotions than this. But communication isn't all about sounds. Even with people, a considerable portion is non-verbal, which means it's transmitted by means of facial expressions and gestures. Depending on which studies you want to believe, the verbal content of a conversation might convey as little as 7 per cent of its actual meaning.[91]

So what about animals? Like us, ravens don't rely on sounds alone. Researchers working with Simone Pika at the Max Planck Institute for Ornithology in Seewiesen discovered that these intelligent birds use their beaks in much the same way we use our hands. Whereas we point to something with a finger, or raise a hand and wave in order to attract someone else's attention to an object or to ourselves, ravens hold up objects with their beak. They also use their beak to point to something or to attract the attention of a member of the opposite sex. With an extensive vocal 'vocabulary' and a number of physical movements that they can adjust to suit the situation in hand, they have an enormously detailed ability to express themselves.[92] They need this because ravens spend almost all their lives together, and they require a way to thoroughly check each other out.

This discovery, however, opens just one small window onto the emotional life of these black birds, which still have many surprises to offer us. There was another such avian 'interpreter' at our forest lodge. Our children had been given a pair of budgerigars as a gift, and Anton, the male, knew how to draw attention to himself. When he wanted to be fed, he picked up his bowl and dropped it. He had plenty of other toys in his cage, so clearly this gesture was a goal-oriented message that read: 'Please fill this bowl!'

But let's leave gestures and get back to speech. Dogs can not only bark, but can also make a variety of sounds that they use in a general way to express them-selves – or perhaps dogs differentiate more precisely between the sounds, while all we can decode is the general message. We got the impression this might be

the case with our Münsterländer, Maxi. As the years
progressed, we learned whether she wanted to tell us
that she was hungry or bored, or that her water bowl
was empty. And even our horses turned out to be capable
of relatively nuanced vocalisations. I was especially
surprised by equine research coming out of Switzerland.
Most horse owners know that horses communicate a
lot using body language. Even though research into
equine non-verbal communication is more advanced than
corvid research, scientists at ETH Zurich were surprised
to discover that there was more meaning in even seemingly primitive calls than previously suspected. They
discovered that whinnies contain two basic frequencies
and can transmit complex information. The first of the
two basic frequencies indicates whether the whinny is
communicating a positive or a negative emotion. The
second frequency indicates the strength of the emotion.[93]
On the relevant page of the ETH website, you can listen
to an example of a whinny for both situations.[94]

After listening to the recordings, I knew right away
that our horses clearly express a positive emotion when
they whinny as they see us coming. True, there's usually
food involved, but I'm not concerned about that. What
interests me is that I can now say with absolute confidence that the horses are happy to see us, something I
used to have to assume. And after reading the research
results, I listened more closely to find out whether their
calls varied, and whether they were happier to see us
at some times than at others. Now I know. Of course
there are times when they are happier, just like people.

Aside from this study, I am convinced that there are also whinnies that express affection. When our older mare, Zipy, cuddles with us, she makes soft, high-pitched sounds through closed lips. When she does that, we know she's feeling good and is happy to be with us. In other words, she's communicating her emotions to us 'verbally'. I think horses are a good example of how little we know about communication among animals. People have kept horses for thousands of years. Therefore in theory they should have been much more thoroughly researched than wild animals. That recent research still manages to surprise us so much makes me even more careful about passing judgement on the skills other species possess.

The next step in communication would be if we not only deciphered the language that animals use to talk to each other, but if we could also talk to them. Then it would be possible to ask them directly about a wide range of emotions, and we could spare ourselves boring scientific studies. And something like this exists already. There is a female gorilla called Koko who has very moving things to say. Yes, I mean 'to say', and she does so using sign language. Penny Patterson trained the young ape as part of her doctoral studies at Stanford University in California. Over time, Koko has learned more than 1,000 signs, and she can understand more than 2,000 words in English. Thanks to her proficiency in sign language, she can let scientists know what she is thinking, and for the first time it is possible to hold an extended conversation with an animal. Other apes have been trained with similar results, indicating that Koko is not an exception.[95] The female gorilla makes regular media appearances and there are often touching scenes. One time she was given a stuffed zebra as a gift. When she was asked what it was, she replied with the signs for 'white' and 'tiger'. And when she was asked where animals go when they die, she signed 'comfortable hole'.[96] Koko has given so many intelligent answers – and has combined concepts she has mastered with new ones – that it really does make sense to call her an ape with a gift for language.

However, there are also strident critics of the Gorilla Foundation, the organisation dedicated to these great apes, whose most important project is exploring Koko's world. An external review of the results isn't possible, as the project publishes very little. Moreover, the conversations with Koko are not conducted according to scientific protocol. For instance, the gorilla often makes mistakes when she answers questions, which the researchers put down to her playful nature.[97] Unfortunately, I cannot tell you what is true and what isn't, from what has been made public, but a gut feeling tells me that at the very least we have severely under-estimated the capacity of our fellow creatures to commu-nicate. And the big question for me is not whether Koko can really speak or whether only some of her answers make sense, because communication between people and animals will always be very one-sided. People try to teach other species human language. The species is then thought to be particularly intelligent when its members understand a lot of concepts or commands and can perhaps even utter a few intelligible words. People are thrilled when budgerigars, ravens or apes like Koko can answer a question in our language.

If we really are the most intelligent species on Earth – and I believe we are – why didn't science approach this from the opposite direction a long time ago? Why are years spent painstakingly teaching lab animals sign language, if modern researchers believe that their capacity to learn is less than ours? Wouldn't it be much easier if we finally began to learn the language of animals? We have many more opportunities now than we had a few years ago, when it would not have been possible to produce sounds at, say, horse level because we lacked the ability to whinny at two different frequencies. Today a computer could do a reasonable job of translating what we want to say into the appropriate animal vocabulary. Unfortunately, I don't know of any serious attempts to do this. There are certainly people who can imitate animal voices – for example, the calls of different species of birds. However, people who can imitate a blackbird or a chickadee can do no more than pipe 'This spot's already taken' in bird language. That's all that the beautiful song trilled by the males sitting up there in the treetops means. What sounds so delightful to our ears serves within the species to scare off competitors. Imagine a parrot that keeps saying, 'Go away.' Unfortunately, that's as far as we've got with our ability to communicate with our fellow creatures.

Where Is the Soul?

So now it's time to dig deep: do animals have a soul, in the sense of an entity with no physical form? It's a really tricky question and one that I would first like to explore in humans because that's a bit easier. So what do we mean by soul? The dictionary, interestingly enough, has a number of different definitions, which shows that there is no common understanding of what the word 'soul' means. One definition suggests the soul is 'the principle of life, feeling, thought and action in humans'. Another suggests it is the spiritual part of people, which, according to religious beliefs, lives on after death.[98] And because no one can inspect the latter, I'm going to set my sights on the first definition.

Feelings, thoughts, and actions – there you have all you need to describe the essence of an animal, right? Animals definitely act and we no longer deny them feelings, so that just leaves thoughts. According to the dictionary definition (which only applies to humans, of course) the capacity to think is a prerequisite for a soul. Okay, let's search for this capacity. That's not easy, though, because there are also a lot of different descriptions for thought, many of which are highly complicated and yet still don't adequately capture the concept. The Technical University of Dresden offered its students this explanation, among others: 'Thinking = a mental process, in which symbolic or pictorial representations of objects, events or actions are generated, transformed or combined.' A much simpler explanation mentioned in the same context puts it more succinctly: 'Thinking is problem solving.'[99] According to this definition, thinking is part of the skill set at least of those animals whose behaviour makes sense to us: ravens that greet other ravens by name, rats that reflect on their actions and regret what they have done, roosters that lie to their hens, and magpies that risk a bit on the side. Would anyone deny that problem-solving processes are going on in their heads?

And with that I would like to come back to the second, religious definition of a soul, after all. Even if I have to tread carefully and I don't feel confident here, even if belief and logic are to a certain extent mutually exclusive, I would like to argue for the existence of an animal soul in the religious sense of the word. If we exclude the possibility of actual physical resurrection, a soul is a prerequisite for life after death. And if you believe that people have souls in this sense, then animals must have them as well. Why? Because the fundamental question is how long people have been going to heaven. For 2,000 years? For 4,000 years? Or for as long as there have been humans? That would be about 200,000 years. But where is the break with earlier life forms, with our predecessors? That break didn't happen abruptly. It happened stealthily. Small changes spread out over generations in the course of evolution. Which individuals are then to be identified as being not yet people with souls? Some female ancestor who lived 200,023 years ago? Or a man armed with a flint weapon who lived 200,197 years ago? No, there is no sharp demarcation, and so you can trace the lineage farther and farther back — past our primitive forefathers, primates, the first mammals, dinosaurs, fish, plants, bacteria. If there is no specific point in time, X, to which the genesis of a creature of the species *Homo sapiens* can be fixed, then there is also no specific point in time when the soul appeared. And if there is a higher form of justice in the religious sense, then when it comes to the question of eternal life, there can't very well be a sharp line drawn between two generations, where the older one is denied entry and the younger one is admitted. Isn't it a beautiful vision that up in heaven there will be a throng of animals of different species living among countless humans?

I personally do not believe in life after death. I envy anyone who does, but my powers of imagination cannot carry me that far. Therefore I am content with the first description of the soul, which I am happy to attribute to all animals. I simply find it a fine thought that other species are not mere machines in which everything is directed by predetermined mechanisms and actions happen when a button is pushed, which is to say when a hormone is released. Squirrels, deer and wild boar with souls: that's the thought that makes life special and warms my heart when I have the opportunity to watch animals like these in the wild.

Epilogue

WHEN I LOOK AT ANIMALS, I like to make analogies to people, because I cannot imagine that they feel so very differently from us. There's a good chance I'm right. The idea that there was an abrupt break in the course of evolution, and that at some point everything was reinvented, is an idea whose time is past. The only major point of contention today is whether animals can think; that's what we do best, after all.

However, what is so important for us could be less important for our fellow creatures – otherwise they would have developed as we did. Is profound thought something that is absolutely necessary? It's certainly not necessary for a satisfying, peaceful life. When we're relaxing on holiday, what's going through our mind is: 'I feel great and I don't have to think about anything at all.' We can experience joy and peace without giving anything much thought, and that is the crux of the matter: emotions have no need for intelligence. As I have stressed, emotions steer instinctive programming and are therefore vital for all species, and therefore all species experience them to a greater or lesser degree. Whether an animal reflects on these emotions, prolongs them through reflection or relives them later is less important. Of course it's nice that we can do exactly this, and perhaps by doing so we get to experience these moments in our lives more intensely. Score one for us. Admittedly that works for less enjoyable moments as well, so score one for the animals, which makes us even.

Why is there still so much resistance to the idea that our fellow creatures have the capacity to feel joy and to suffer? This resistance comes from some scientists, but above all from politicians who answer to farmers. Mostly they are protecting the cheap methods used by factory farms to house and handle animals, such as castrating piglets without anaesthetic, as I mentioned earlier. And then there's hunting, which claims the lives of hundreds of thousands of large mammals and many birds every year, and which in its current form is simply no longer appropriate.

When all the arguments have been made and it's clear we're getting to the point where we must grant animals way more skills than we usually do, the knockout punch is delivered: the charge of anthropomorphism. People who compare animals to humans, so the argument goes, are unscientific. They are wishful – maybe even mystical – thinkers. In the heat of the fray, an essential truth we all learned in school is overlooked: a human being is, from a purely biological perspective, an animal and therefore not so very different from other species. It follows that a comparison between people and animals is not too much of a stretch, especially since we can only relate to and empathise with things we understand. And so it makes sense to take a closer look at those animals in which we can detect emotions and mental processes similar to our own. This comparison is easier with feelings such as hunger and thirst, whereas comparing human and animal experiences of joy, grief or compassion makes some people's hair stand on end.

The goal is not to make animals seem like us, but to help us understand them better. Most importantly, these comparisons serve to point out that animals are not dim-witted creatures clearly stuck a level below us on the evolutionary scale, creatures that experience only pale imitations of our rich range of sensations for pain and other such feelings. No. People who understand that deer, wild boar and ravens lead their own lives, perfect in their own way, and have a lot of fun while they're at it might even respect animals as insignificant as the tiny weevils that rummage around, contented and happy, in the leaf litter of ancient forests.

One reason there is still doubt about the emotional world of animals might be that many emotions and mental processes are not yet clearly defined even in people. In this context, let's remember what we said about happiness, gratitude or even just thinking – all terms that have been difficult to describe. How can we understand something in animals that we can't even clearly grasp in ourselves? Pure science, which today is defined by its demand for objectivity, might not help us advance, because it requires us to set aside our emotions. However, because people are mostly driven by their emotions, as we saw in the context of instincts, we possess the appropriate antennae to recognise stirrings of emotion in others. And why should these antennae fail, just because the other in question is an animal and not a person?

We evolved in a world full of other species, and we had to survive despite them and with them. It was surely just as important to be able to read the intentions of wolves, bears or wild horses as it was to read the faces of strangers. No doubt our senses can sometimes deceive us and we can read too much into the behaviour of dogs and cats. But I'm convinced that in the majority of cases our intuitions are correct. Current scientific discoveries come as no real surprise to animal lovers. All they do is give us more confidence to trust our own feelings where animals are concerned.

When people reject acknowledging too much in the way of emotions in animals, I have the vague feeling that there's a bit of fear that human beings could lose their special status. Even worse, it would become much more difficult to exploit animals. Every meal eaten or leather jacket worn would trigger moral considerations that would spoil their enjoyment. When you think how sensitive pigs are, how they teach their young and help them deliver their own children later in life, how they answer to their names and pass the mirror test, the thought of the annual slaughter of 250 million of these animals across the European Union alone is chilling.[100]

And it doesn't stop with animals. As science has discovered and you might well already have read, we must now acknowledge that trees and other plants have feelings and even a capacity to remember. How, then, are we supposed to feed ourselves in a morally acceptable manner if we are now justified in feeling sorry for plants, too? Like many species, we cannot photosynthesise to create our own food, so we have to eat living entities to survive. The choices we make are very personal. They might depend on where we live or the culture in which we were brought up. Ultimately, though, each of us has to decide what we will eat. My hope is that what you have learned in this book will help you make informed decisions for the future.

From my personal perspective, I am suggesting that we infuse our dealings with the living beings with which we share our world with a little more respect, as we once used to do, whether those beings are animals or plants. That doesn't mean completely doing without them, but it does mean a certain reduction in our level of comfort and in the amount of biological goods we consume. As a reward, if we then have happier horses, goats, chickens and pigs; if we can then observe contented deer, martens or ravens; if one day we can listen in when the ravens call their names, then a hormone will be released into our central nervous systems that will spread a feeling against which we have no defence – happiness!

Acknowledgements

I OWE MY WIFE, MIRIAM, a huge thank you. Once again she worked through my unfinished manuscript many times, reviewing the thoughts I had committed to paper with a critical eye. My children, Carina and Tobias, helped jog my memory as I brooded yet again in front of a blank screen, completely unable to come up with a fitting anecdote no matter how I tried, even though there were so many to choose from – thank you both, my dears! The team at Ludwig Verlag, my publisher, came up with the idea for a book centred on animals (ah yes, there were so many ideas swimming around in my head that I could have written three books). Thank you!

Angelika Lieke gave the text one final polish, improving its readability by pointing out repetitions, illogical sentences and impediments to understanding. I don't want to forget my agent, Lars Schultze-Kossack, who made the connection with the publisher and gave me constant encouragement when I was harbouring sincere doubts about whether this could ever turn into a book (as he did with my book about the hidden life of trees, when I was equally unsure of myself). And last but not least, I want to thank Maxi, Schwänli, Vito, Zipy, Bridgi and all the other four-legged and two-winged helpers that have allowed me to be a part of their rich lives. In the end, they were the ones that had all the stories to tell, which, dear reader, I then undertook to translate for you.

Notes

1. Max Planck Institute for Human Cognitive and Brain Sciences, 'Unconscious Decisions in the Brain', 14 April 2008, www.mpg.de/research/unconscious-decisions-in-the-brain

2. McGill Newsroom, 'Squirrels Show Softer Side by Adopting Orphans', news release, 1 June 2010, www.mcgill.ca/newsroom/channels/news/squirrels-show-softer-side-adopting-orphans-163790

3. *The Guardian*, 'French Bulldog Called Baby Adopts Six Wild Boar Piglets', 15 February 2012, www.theguardian.com/world/2012/feb/15/french-bulldog-wild-boar-piglets

4. *Toronto Star*, 'Yeti the Farm Dog Nurses 14 Piglets in Cuba', 3 September 2011, www.thestar.com/news/world/2011/09/03/yeti_the_farm_dog_nurses_14_piglets_in_cuba.html

5. Amy DeMelia, 'The Tale of Cassie and Moses', *Sun Chronicle*, 5 September 2011, www.thesunchronicle.com/news/the-tale-of-cassie-and-moses/article_e9d792d1-c55a-51cf-9739-9593d39a8ba2.html

6. Von Antje Joel, 'Mit diesem Delfin stimmt etwas nicht' (Something's wrong with this dolphin), *Die Welt*, 26 December 2011, www.welt.de/wissenschaft/umwelt/article13782386/Mit-diesem-Delfin-stimmt-etwas-nicht.html

7. Dr C. George Boeree, 'The Emotional Nervous System', Shippensburg University Webspace, 2009, http://webspace.ship.edu/cgboer/limbicsystem.html

8. Victoria Braithwaite, *Do Fish Feel Pain?*, Oxford: Oxford University Press, 2010

9. J.S. Feinstein et al., 'The Human Amygdala and the Induction and Experience of Fear', *Current Biology* 21, no. 1 (11 January 2011): 34–8; doi: 10.1016/j.cub.2010.11.042

10. M. Portavella García et al., 'Avoidance Response in Goldfish: Emotional and Temporal Involvement of Medial and Lateral

Telencephalic Pallium', *The Journal of Neuroscience* 24, no. 9 (3 March 2004): 2335–42; doi: 10.1523/JNEUROSCI.4930–03.2004

11. H. Breuer, 'Die Welt aus der Sicht einer Fliege' (A fly's eye view of the world), *Süddeutsche Zeitung*, 19 May 2010, http://www.suedeutsche.de/panorama/forschung-die-welt-aus-sicht-einer-fliege-1.908384

12. Forschungsverbund Berlin e.V. (FVB), 'Do Fish Feel Pain? Not as Humans Do, Study Suggests', news release, *ScienceDaily*, www.sciencedaily.com/releases/2013/08/130808123719.htm; *Spiegel Online*, 'Fische kennen keinen Schmerz wie wir' (Fish don't feel pain like we do), 9 September 2013, http://www.spiegel.de/wissenschaft/natur/angelprofessor-robert-arlinghaus-ueber-den-schmerz-der-fische-a-920546.html

13. M. Evers, 'Leiser Tod im Topf' (A quiet death in the pot), *Der Spiegel*, 52 (2015): 120

14. T. Stelling, 'Do Lobsters and Other Invertebrates Feel Pain? New Research Has Some Answers', *The Washington Post*, 3 October 2014, https://www.washingtonpost.com/national/health-science/do-lobsters-and-other-invertebrates-feel-pain-new-research-has-some-answers/2014/03/07/f026ea9e-9e59-11e3-b8d8-94577ff66b28_story.html

15. J. Dugas-Ford et al., 'Cell-Type Homologies and the Origins of the Neocortex', *Proceedings of the National Academy of Sciences* 109, no. 42 (October 2012): 16974–9; doi: 10.1073/pnas.1204773109

16. C.R. Reid et al., 'Slime Mold Uses an Externalized Spatial "Memory" to Navigate Complex Environments', *PNAS* 109, no. 43: 17490–4; doi: 10.1073/pnas.1215037109

17. American Association for the Advancement of Science, 'Slime Design Mimics Tokyo's Rail System: Efficient Methods of a Slime Mold Could Inform Human Engineers', news release, *Science Daily*, 22 January 2010, https://www.sciencedaily.com/releases/2010/01/100121141051.htm

18. Washington State Recreation and Conservation Office, Washington Invasive Species Council, 'Feral Swine', http://www.invasivespecies.wa.gov/priorities/feral_swine.shtml

19. Kirstin Lauterbach et al., 'Do All Male Wild Boar Yearlings *Sus Scrofa L.* Leave Home?', Sixth International Symposium

on Wild Boar (*Sus scrofa*) and Sub-Order Suiformes, Kykkos, Cyprus, October 2006; doi: 10.13140/2.1.5146

20. Statista, 'Jahresstrecken von Schwarzwild (Wildschweine) in Deutschland von 1997/98 bis 2014/15' (Annual numbers of wild boar in Germany from 1997/8 to 2014/15), http://de.statista.com/statistik/daten/studie/157728/umfrage/jahresstrecken-von-schwarzwild-in-deutschland-seit-1997-98/

21. E. Bodderas, 'Schweine sprechen ihre eigene Sprache. Und bellen' (Pigs speak their own language. And bark), *Die Welt*, 15 January 2012, http://www.welt.de/wissenschaft/article13813590/Schweine-sprechen-ihre-eigene-Sprache-Und-bellen.html

22. Solon Kelleher, 'Tangled Whale Gives His Rescuers the Best Thank You Ever', *the dodo* (blog), 8 July 2015, www.thedodo.com/tangled-whale-says-thank-you-1238266959.html

23. Katy Sewall, 'The Girl Who Gets Gifts from Birds', *BBC News*, 25 February 2015, http://www.bbc.com/news/magazine-31604026

24. Anders Pape Moller, 'Deceptive Use of Alarm Calls by Male Swallows, *Hirundo rustica*: A New Paternity Guard', *Behavioral Ecology* 1, no. 1 (June 1990): 1–6; doi: 10.1093/beheco/1.1.1

25. Michael Becker, 'Elstern' (Magpies), March 1999, http://www.ijon.de/elster/verhalt.html

26. Rebecca Grambo, *The Nature of Foxes*, Vancouver: Greystone, 1995, p. 30

27. Rachael C. Shaw and Nicola S. Clayton, 'Careful Cachers and Prying Pilferers: Eurasian Jays *(Garrulus glandarius)* Limit Auditory Information Available to Competitors', *Proceedings of the Royal Society B*, 5 December 2012; doi: 10.1098/rspb.2012.2238

28. Max Planck Institute for Ornithology, 'Competition Favours Shy Tits', 10 March 2016, http://www.orn.mpg.de/3685340/news_publication_10363050

29. C. Turbill et al., 'Regulation of Heart Rate and Rumen Temperature in Red Deer: Effects of Season and Food Intake', *Journal of Experimental Biology* 214, no. 6 (2011): 963–70; doi: 10.1242/jeb.052282

30. University of Veterinary Medicine, Vienna, 'Personality Differences: In Lean Times Red Deer with Dominant Personalities Pay a High Price', news release, 18 September

2013, www.vetmeduni.ac.at/en/infoservice/presseinforma-tion/press-releases-2013/press-release-09-18-2013-personality-differences-in-lean-times-red-deer-with-dominant-personalities-pay-a-high-price

31. Freie Universität Berlin, 'When Bees Can't Find Their Way Home', news release, 3 March 2014, http://www.fu-berlin.de/en/presse/informationen/fup/2014/fup_14_092-bienenorientierung-pestizide-publikation-menzel/index.html

32. S. Klein, 'Die Biene weiss, wer sie ist' (The bee knows who it is), *Zeit Magazin*, 25 February 2015, http://www.zeit.de/zeit-magazin/2015/02/bienen-forschung-randolf-menzel; Randolf Menzel et al., 'A Common Frame of Reference for Learned and Communicated Vectors in Honeybee Navigation', *Current Biology* 21, no. 8 (26 April 2011): 645–50; doi: 10.1016/j.cub.2011.02.039

33. Alan Bellows, 'Clever Hans, the Math Horse', *Damn Interesting*, article 168, https://www.damninteresting.com/clever-hans-the-math-horse/

34. A. Lebert and C. Wüstenhagen, 'In Gedanken bei den Vögeln' (In the minds of birds), *Zeit Online*, 4/2015, http://www.zeit.de/zeit-wissen/2015/04/hirnforschung-tauben-onur-guentuerkuen; Lorenzo von Fersen and Juan Delius, 'Long-Term Retention of Many Visual Patterns by Pigeons', *Ethology* 82, no. 2 (January 2010): 141–55; doi: 10.1111/j.1439-0310.1989.tb00495.x

35. Aleksey Vnukov, 'Crowboarding: Russian Roof-Surfin' Bird Caught on Tape', *YouTube*, 12 January 2012, https://www.youtube.com/watch?v=3dWw9GLcOeA

36. Wikipedia, s.v. 'Marriage', https://en.wikipedia.org/wiki/Marriage

37. Anne Jeschke, 'Zu welchen Gefühlen Tiere wirklich fähig sind' (What are animals really capable of feeling?), *Die Welt*, 15 February 2015, https://www.welt.de/wissenschaft/umwelt/article137478255/Zu-welchen-Gefuehlen-Tiere-wirklich-faehig-sind.html

38. Herbert Cerutti, 'Clevere Jagdgefährten' (Clever hunting associates), *NZZ Folio*, July 2003, http://folio.nzz.ch/2003/juli/clevere-jagdgefahrten

39. Sindya N. Bhanoo, 'Ravens Can Recognize Old Friends, and Foes, Too', *New York Times*, 23 April 2012, http://www.

nytimes.com/2012/04/24/science/ravens-can-recognize-old-friends-and-foes-too.html

40. J. Kirchner, G. Manteuffel and L. Schrader, 'Individual Calling to the Feeding Station Can Reduce Agonistic Interactions and Lesions in Group Housed Sows', *Journal of Animal Science* 90, no. 13 (December 2012): 5013–20, https://www.animalscience-publications.org/publications/jas/articles/90/13/5013

41. Jeremy Hance, 'Birds Are More Like "Feathered Apes" Than "Bird Brains"', *The Guardian*, 5 November 2016, https://www.theguardian.com/environment/radical-conservation/2016/nov/05/birds-intelligence-tools-crows-parrots-conservation-ethics-chickens

42. Donald M. Broom, Hilana Sena and Kiera L. Moynihan, 'Pigs Learn What a Mirror Image Represents and Use It To Obtain Information', *Animal Behavior* 78, no. 5 (November 2009): 1037–41; doi: 10.1016/j.anbehav.2009.07.027

43. McGill Newsroom, 'Squirrels Show Softer Side by Adopting Orphans', news release, 1 June 2010, https://www.mcgill.ca/newsroom/channels/news/squirrels-show-softer-side-adopting-orphans-163790

44. Discovery of Sound in the Sea, 'How Do Marine Fish Communicate Using Sound?', http://www.dosits.org/animals/useofsound/fishcommunicate/

45. Mathias Kneppler, 'Auswirkung des Forst- und Alpwegebaus im Gebirge auf das dort lebende Schalenwild und seine Bejagbarkeit' (Effects of forest and mountain roads in the mountains on the ungulates that live there and hunting them), Abschlussarbeit des Universitätslehrgangs Jagdwirt/-in an der Universität für Bodenkultur Wien Lehrgang VI (final thesis for the course in game management at the University of Natural Resources and Life Sciences, Vienna, Course VI), 2013/14, p. 7

46. S. Herrmann, 'Peinlich' (Embarrassing), *Süddeutsche Zeitung*, 30 May 2008, http://www.sueddeutsche.de/wissen/schamgefuehle-peinlich-1.830530; Christine R. Harris, 'Embarrassment: A Form of Social Pain', *American Scientist* 94, no. 6 (November–December 2006): 524–33, http://charris.ucsd.edu/articles/Harris_AS2006.pdf

47. Mary Bates, 'Rats Regret Making the Wrong Decision', *Wired*, 8 June 2014, https://www.wired.com/2014/06/rats-regret-making-the-wrong-decision/

48. CBS DFW, 'Scold Them All You Want, Dogs Feel No Shame Says Behaviorist', 26 February 2014, http://dfw.cbslocal. com/2014/02/26/scold-them-all-you-want-dogs-feel-no-shame-says-behaviorist/

49. Elsevier, 'What Really Prompts the Dog's "Guilty Look"', *ScienceDaily*, 14 June 2009, https://www.sciencedaily.com/releases/2009/06/090611065839.htm

50. Friederike Range, Lisa Horna, Zsófia Viranyib and Ludwig Hubera, 'The Absence of Reward Induces Inequity Aversion in Dogs', *PNAS* 106, no. 1 (6 January 2009): 340–5. Reported by Nell Greenfieldboyce in 'Dogs Understand Fairness, Get Jealous, Study Finds', *NPR*, 9 December 2008, http://www.npr.org/templates/story/story.php?storyId=97944783

51. Jorg J.M. Massen, Caroline Ritter and Thomas Bugnyar, 'Tolerance and Reward Equity Predict Cooperation in Ravens (*Corvus corax*)', *Scientific Reports* 5, no. 15021 (2015); doi: 10.1038/srep15021

52. Ishani Ganguli, 'Mice Show Evidence of Empathy', *The Scientist*, 30 June 2006, http://www.the-scientist.com/?articles.view/articleNo/24101/title/Mice-show-evidence-of-empathy/

53. J. Martin Loren et al., 'Reducing Social Stress Elicits Emotional Contagion of Pain in Mouse and Human Strangers', *Current Biology* 25, no. 3 (2 February 2015): 326–32; doi: 10.1016/j.cub.2014.11.028

54. Felicity Muth, 'Can Pigs Empathize?', *Scientific American*, 13 January 2015, https://blogs.scientificamerican.com/not-bad-science/can-pigs-empathize/

55. Aleksander Medveš, 'Crow Rescue', *YouTube*, 27 April 2016, https://www.youtube.com/watch?v=Nubc09jTW-M

56. Mark Matousek, 'The Meeting Eyes of Love: How Empathy Is Born in Us', 8 April 2011, https://www.psychologytoday.com/blog/ethical-wisdom/201104/the-meeting-eyes-love-how-empathy-is-born-in-us

57. Henry H. Lee, Michael N. Molla, Charles R. Cantor and James J. Collins, 'Bacterial Charity Work Leads to Population-Wide Resistance', *Nature* 467 (2 September 2010): 82–5; doi: 10.1038/nature09354

58. G.G. Carter and G.S. Wilkinson, '2013 Food Sharing in Vampire Bats: Reciprocal Help Predicts Donations More than Relatedness

or Harassment', *Proceedings of the Royal Society B* (2 January 2013); doi: 10.1098/rspb.2012.257

59. Birk Grüling, 'Ein Hund im Wolfspelz ist Tierquälerei' (A dog in wolf's clothing is animal abuse), *Zeit Online*, 3 July 2014, http://www.zeit.de/wissen/umwelt/2014-06/tierhaltung-wolf-hybrid-hund; Patricia McConnell, 'The Tragedy of Wolf Dogs', *The Other End of the Leash* (blog), 13 July 2013, http://www.patriciamcconnell.com/theotherendoftheleash/the-tragedy-of-wolf-dogs

60. University of Massachusetts Amherst, 'Lord's Study may Explain why Wolves are Forever Wild, but Dogs can be Tamed', news release, 17 January 2013, https://www.umass.edu/newsoffice/article/lords-study-may-explain-why-wolves-are-forever-wild-dogs-can-be-tamed

61. 'Rehbock greift zwei Frauen beim Walking an' (Buck attacks two women out walking), *Schwarzwälder Bote*, 21 April 2015, http://www.schwarzwaelder-bote.de/inhalt.st-georgen-rehbock-greift-zwei-frauen-beim-walking-an.aee11194-f2cd-40b8-ba43-5204586dfc0c.html

62. Dana Krempels, 'The Mystery of Rabbit Poop', Miami College of Arts & Sciences, http://www.bio.miami.edu/hare/poop.html

63. Jennifer Welsh, 'Why Pooh Bear Loves Honey, But Tigger Doesn't', *Live Science*, 12 March 2012, http://www.livescience.com/18994-carnivores-taste-sweets.html

64. Katsuhisa Ozaki et al., 'A Gustatory Receptor Involved in Host Plant Recognition for Oviposition of a Swallowtail Butterfly', *Nature Communications* 2, no. 42 (2011); doi: 10.1038/ncomms1548

65. Patricia McConnell, 'The Other End of the Leash: Why Do Dogs Roll in Disgusting Stuff?', *The Other End of the Leash* (blog), 1 June 2015, http://www.patriciamcconnell.com/theotherendoftheleash/why-do-dogs-roll-in-disgusting-stuff

66. L.A. Smith, P.C.L. White and M.R. Hutchings, 'Effect of the Nutritional Environment and Reproductive Investment on Herbivore–Parasite Interactions in Grazing Environments', *Behavioral Ecology* 17, no. 4 (28 April 2006): 591–6; doi: 10.1093/beheco/ark004

67. Jenny Stanton and Jane Flanagan, 'Beware, Lions Crossing (and Mauling)!', *MailOnline*, 13 July 2015, http://www.dailymail.

co.uk/news/article-3159262/Beware-lions-crossing-mauling-Breathtaking-moment-beasts-catch-antelope-inches-stunned-tourists-safari-park-road.html

68. Dr Michael Petrak, 'Rotwild als erlebbares Wildtier – Folgerungen aus dem Pilotprojekt Monschau-Elsenborn für den Nationalpark Eifel' (Red deer as watchable wildlife – results of a pilot project in Monschau-Elsenborn for Eifel National Park), in *Von der Jagd zur Wildbestandsregulierung*, NUA-Heft no. 15, Natur-und Weltschutz-Akademie des Landes Nordrhein-Westfalen (NUA), May 2004, p. 19

69. Association for Psychological Science, 'The Genetics of Fear', news release, 9 March 2009, http://www.psychologicalscience. org/news/releases/the-genetics-of-fear-study-suggests-specific-genetic-variations-contribute-to-anxiety-disorders. html

70. Dietmar Spengler, 'Genes Learn from Stress', Research Report 2010, Max Planck Institute of Psychiatry, https://www.mpg. de/431776/forschungsSchwerpunkt?c=148053

71. Amy Liptrot, 'How Berlin's Urban Goshawks Helped Me Learn to Love the City', *The Guardian*, 13 May 2005, https://www. theguardian.com/cities/2015/may/13/berlin-goshawks-urban-wildlife-tempelhof-airport-birdwatching

72. Kathryn Westcott, 'What Is Stockholm Syndrome?', *BBC News Magazine*, 22 August 2013, http://www.bbc.com/news/magazine-22447726

73. Dietrich von Holst, 'Populationsbiologische Untersuchungen beim Wildkaninchen' (Research into the population dynamics of wild rabbits), *LÖBF-Mitteilungen* (Transactions of the Institute of Ecology, Land Use and Development, and Forestry, North Rhine-Westphalia), no. 1, 2004, p. 17

74. Canadian Honey Council, 'How to Make a Pound of Honey', http://www.honeycouncil.ca/chc_poundofhoney.php

75. Nick Jans, *The Grizzly Maze: Timothy Treadwell's Fatal Obsession with Alaskan Bears*, New York: Dutton, 2005

76. Eva Bellemain et al., 'The Dilemma of Female Mate Selection in the Brown Bear, a Species with Sexually Selected Infanticide', *Proceedings of the Royal Society B* 273(1584), (7 February 2006): 283–91; doi: 10.1098/rspb.2005.3331

77. MIT News, 'Rats Dream about Their Tasks during Slow Wave
     Sleep', 18 May 2002, http://news.mit.edu/2002/dreams

78. M. Jouvet, 'States of Sleep', *Scientific American*, 1 February
     1967, https://www.scientificamerican.com/article/the-states-
     of-sleep/

79. H. Breuer, 'Die Welt aus der Sicht einer Fliege' (A fly's eye
     view of the world), *Süddeutsche Zeitung*, 19 May 2010, http://
     www.suedeutsche.de/panorama/forschung-die-welt-aus-sicht-einer-
     fliege-1.908384

80. Elke Maier, 'A Four-Legged Early-Warning System', Max
     Planck Institutes, https://www.mpg.de/8252362/W004_
     Environment-Climate_058-063.pdf

81. G. Berberich and U. Schreiber, 'GeoBioScience: Red Wood Ants
     as Bioindicators for Active Tectonic Fault Systems in the West
     Eifel (Germany)', *Animals (Basel)* 3, no. 2 (17 May 2013):
     475–98; doi: 10.3390/ani3020475

82. K.M. Williams et al., 'Increasing Prevalence of Myopia in
     Europe and the Impact of Education', *Opthamology* 122, no. 7
     (July 2015): 1489–97; doi: 10.1016/j.ophtha.2015.03.018

83. Emily Benson, '"Landmark Study" Solves Mystery behind
     Classic Evolution Story', *Science*, 1 June 2016, http://www.
     sciencemag.org/news/2016/06/landmark-study-solves-mystery-
     behind-classic-evolution-story

84. A. Lebert and C. Wüstenhagen, 'In Gedanken bei den Vögel'
     (In the minds of birds), *Zeit Online*, 4/2015, http://www.zeit.
     de/zeit-wissen/2015/04/hirnforschung-tauben-onur-guen-
     tuerkuen; J.D. Delius, R.J Perchard and J. Emmerton, 'Polarized
     Light Discrimination by Pigeons and an Electroretinographic
     Correlate', *Journal of Comparative Physiological Psychology* 90,
     no. 6 (June 1976): 560–71, https://www.ncbi.nlm.nih.gov/
     pubmed/956468

85. Paul E. Miller and Christopher J. Murphy, 'Vision in Dogs',
     Journal of the American Veterinary Medical Association 207, no.
     12 (15 December 1995), http://redwood.berkeley.edu/bruno/
     animal-eyes/dog-vision-miller-murphy.pdf

86. Sarah Bordenstein, 'What Is a Tardigrade?', *Microbial Life
     Educational Resources*, http://serc.carleton.edu/microbelife/
     topics/tardigrade/index.html

87. *National Geographic*, 'The Human Condition: Our Imprint Deepens as Consumption Accelerates', http://www.national-geographic.com/earthpulse/human-impact.html

88. Gesamtverband der Deutschen Versicherungswirtschaft, 'Zahl der Wildunfälle sinkt auf 247.000' (The number of accidents involving wildlife declines to 247,000), 10 July 2014, http://www.gdv.de/2014/10/zahl-der-wildunfaelle-sinkt-leicht/

89. Peter Werner and Rudoph Zahner, 'Biological Diversity and Cities: A Review and Bibliography', Bundesamt für Naturschutz, *BfN-Skripten* 245 (2009)

90. Michael Slezak, 'Ancient DNA Suggests Dogs Split from Wolves 40,000 Years Ago', *New Scientist*, 27 May 2015, https://www.newscientist.com/article/mg22630235-500-ancient-dna-suggests-dogs-split-from-wolves-40000-years-ago/

91. Stefan Müller, 'Interkulturelles Marketing' (Intercultural marketing), PowerPoint presentation, 2007, slide 4, 'Paraverbale Kommunikation' (Paraverbal communication), https://tu-dresden.de/gsw/wirtschaft/marketing/ressourcen/dateien/lehre/lehre_pdfs/Mueller_IM_G1_Kommunikation.pdf?lang=de; Jeff Thompson, 'Is Nonverbal Communication a Numbers Game?', *Psychology Today*, 30 September 2011, https://www.psychologytoday.com/blog/beyond-words/201109/is-nonverbal-communication-numbers-game

92. Max Planck Institute, '"Look at That!" Ravens Use Gestures, Too', 29 November 2011, https://www.mpg.de/4664902/ravens_use_gestures

93. E.F. Briefer et al., 'Segregation of Information about Emotional Arousal and Valence in Horse Whinnies', *Scientific Reports* 4, no. 9989 (2015); doi: 10.1038/srep09989

94. Inken De Wit, 'Wie Pferde Emotionen äussern' (How horses express emotion), ETH Zürich, https://www.ethz.ch/de/news-und-veranstaltungen/eth-news/news/2015/05/wiehern-nicht-gleich-wiehern.html

95. The Gorilla Foundation, http://www.koko.org

96. Danny D. Steinberg, Hiroshi Nagata and David P. Aline, *Psycholinguistics: Language, Mind, and Word*, London: Longman, 1982, p. 150; Roc Morin, 'A Conversation with Koko the Gorilla', *The Atlantic*, 28 August 2015, http://www.theatlantic.

com/technology/archive/2015/08/koko-the-talking-gorilla-sign-language-francine-patterson/402307/

97. J.C. Hu, 'What Do Talking Apes Really Tell Us?', *Slate*, 20 August 2014, http://www.slate.com/articles/health_and_science/science/2014/08/koko_kanzi_and_ape_language_research_criticism_of_working_conditions_and.html

98. Dictionary.com, s.v. 'Soul', http://www.dictionary.com/browse/soul

99. Thomas Goschke, 'Kognitionspsychologie: Denken, Problemlösen, Sprache' (Cognitive psychology: thinking, problem solving, speech), Module A1: Kognitive Prozesse (Cognitive processes), PowerPoint presentation, 2013, https://tu-dresden.de/mn/psychologie/allgpsy/ressourcen/dateien/lehre/lehreveranstaltungen/goschke_lehre/ss2013/folder-2013-04-15-9955666685/vl01_einfuehrung?lang=en

100. Pol Marquer et al., 'Pig Farming Sector – Statistical Portrait 2014', Eurostat, *Statistics Explained*, http://ec.europa.eu/eurostat/statistics-explained/index.php/Pig_farming_sector_-_statistical_portrait_2014#Production_of_porkmeat

THE HIDDEN LIFE OF TREES
What They Feel, How They Communicate
Discoveries from a Secret World

The international bestseller

How do trees live? Do they feel pain, or have awareness of their surroundings?

In *The Hidden Life of Trees*, forester Peter Wohlleben shares his deep love of woods and forests and explains the amazing processes of life, death and regeneration he has observed in the woodland and the amazing scientific processes behind the wonders of which we are blissfully unaware.

Much like human families, tree parents live together with their children, communicate with them and support them as they grow, sharing nutrients with those who are sick or struggling and creating an ecosystem that mitigates the impact of extremes of heat and cold for the whole group. Drawing on groundbreaking new discoveries, Wohlleben presents the science behind the secret and previously unknown life of trees and their communication abilities, and he describes how these discoveries have informed his own practices in the forest around him. After a walk through the woods with Wohlleben, you'll never look at trees in the same way again.

Paperback £9.99 ISBN 978-0-00-821843-0